A PRIMER TO MECHANISM IN ORGANIC CHEMISTRY

A primer to mechanism in organic chemistry

Peter Sykes M.Sc., Ph.D., F.R.S.C., C.Chem
Fellow of Christ's College, Cambridge

 LONGMAN

Addison Wesley Longman Limited
Edinburgh Gate, Harlow
Essex CM20 2JE, England
and Associated Companies throughout the world

First published 1995
Reprinted 1997, 1998
Translations into
Japanese 1995
German 1996
Italian 1997
In Preparation
Greek
Russian

British Library Cataloguing in Publication Data
A catalogue entry for this title is available from the British Library.

ISBN 0-582-26644-0

Library of Congress Cataloging-in-Publication Data
A catalog entry for this title is available from the Library of Congress.

Typeset by 16 in 10 on 12 pt Monotype Times
Produced by Addison Wesley Longman Singapore (Pte) Ltd
Printed in Singapore

For Kate and William with love

Contents*

*A detailed list of contents will be found at the beginning of each chapter.

Foreword

Teaching and learning organic chemistry are not becoming easier. Its scope and subtlety are increasing, while curriculum time is not; and would-be students seem all too easily diverted through a poor public perception of chemistry.

Our experience here shows that it is the introductory part of the subject where the difficulty is greatest, whether approached from the perspective of student or teacher. Arguably that is how things have always been, but the magnitude of the problem seems to have increased in recent years. By and large, once students have acquired some basic experience of the systematic vocabulary and mechanistic underpinning of organic chemistry they seem able to take increasingly advanced material in their stride. Indeed many come to accept, if only in retrospect, that the subject really does get better as it goes along and that the effort spent in acquiring introductory skills will be repaid many times over. Overcoming the initial hurdle remains a problem, however, and teachers stand or fall on the extent to which they can catalyse the vital introductory process: to attract and hold student interest while presenting the subject with a reasonable degree of rigour.

This book is just such a catalyst. Dr. Sykes is a renowned teacher, and one of the most successful textbook authors in the chemical world. The book draws on the experience of a large part of his professional life. It shares with his earlier books a co-ordinating focus on mechanistic principles, but its structure is very different. Conceptually it could hardly be simpler, and it represents an approach, and is written in a style, that students and teachers of elementary organic chemistry can draw upon—in whole or in part—with advantage. I commend it with enthusiasm!

Melbourne, Australia   D. W. CAMERON
December 1994

Preface

It is now well over thirty years since my *Guidebook to Mechanism in Organic Chemistry* first saw the light of day, and during that time a great deal has happened in the teaching of chemistry, and not least in the teaching of organic chemistry. There have been six significant revisions of the *Guidebook* over this period of time and now, despite considerable pressure to undertake a seventh, I have decided that the time has come for a new departure.

There is no doubt that, over this thirty year span, the *Guidebook* has—despite manful efforts to the contrary—become a good deal more sophisticated; not merely in the topics considered, but also in the arguments offered to explain them. I believe that there is now a real need for a simpler book, and have, therefore, decided to go back to square one and start all over again. The *Primer* is a considerably simpler book, one that seeks to set out the basic, underlying framework of organic reaction mechanisms, illustrated—and I hope illumined—by the simplest of examples. This book is **not**, I hasten to add, a "son-of-*Guidebook*"; I have sought to think through the subject matter *de novo*, and the general arrangement is indeed now quite different.

The basic premise of the book is that it is possible—at this level—to make some sense out of the apparent overfacing complexity of organic chemistry on the basis of three underlying axioms: that there are only <u>three</u> types of reaction—substitution, addition and elimination; that these reactions involve only <u>three</u> types of reagent—nucleophiles, electrophiles and radicals; and that there are only <u>two</u> effects—electronic and steric—through which the behaviour of a bond, or group, undergoing reaction, can be influenced by the rest of the molecule.

There is no discussion of bonding that involves orbital theory, nor—in formal terms—of chemical energetics: discussion of either would have taken up more space than their inclusion would have justified in what is intended to be, above all, a short and simple book. Satisfactory explanations can, at this level, be provided without an absolute need for either topic, and their consideration is, I believe, better left till a rather more sophisticated treatment of the subject is possible, e.g. at the level of the *Guidebook*.

This opinion marks a considerable change on my part: thirty years ago I would have regarded it as heretical! In the intervening time, however, I have seen only too often how a student's on-going difficulty with the above topics can actually get in the way of his or her acquiring a more desirable ability: an instinctive "feel" for the likely course taken by organic reactions.

I have been most fortunate with my previous books in that readers have been generous in writing to point out errors, ambiguities and poor explanations: I keep a copy interleaved with blank pages in which all comments are recorded as received, for incorporation in the next reprint. I should be most grateful if readers would be equally generous in their response to this new book.

My warmest thanks are due to Professor H. Hopf of the University of Braunschweig, Germany for his encouragement to go on with the project, and for then–nobly–being prepared to undertake a German translation; to Philip Jones for so kindly inviting me to participate in his annual courses at the American School of Paris, where many of the ideas were first tried out; to Ron Harper of Longman, Australia who suggested incorporating a summary at the end of each chapter; to my old friend Professor K. E. Russell of Queen's University, Kingston, Ontario, Canada and to Dr J. C. Walton of the University of St. Andrews, Scotland, for generous advice on (but no responsibility for!) Chapter 10; to another old friend, Professor D. W. Cameron of the University of Melbourne, Australia, for so kindly agreeing to write a Foreword: a reminder of the great warmth and kindness I have been shown in his most hospitable Department, over the course of many happy visits; finally, to my wife, Joyce, who has as usual made me fight for every word and diagram: without her continuing care and love it just would not have been possible.

Cambridge   PETER SYKES
September 1994

1

Basics

A—perhaps the—major difficulty experienced in seeking an understanding of organic chemistry is the immense number of different compounds that are involved: considerably more than ten million compounds already, and the number growing by many hundred thousand new ones every year!

To try to make some sense out of all this multiplicity, we have to devise something in the way of a unifying principle, or principles, to help guide us through the maze. There have been a number of attempts to do this, of which the simplest, and most familiar, is the idea of **functional groups**.

1.1 FUNCTIONAL GROUPS

The majority of organic compounds contain one or more characteristic atoms, or groups of atoms, generally referred to as functional groups, of which the following are typical examples:

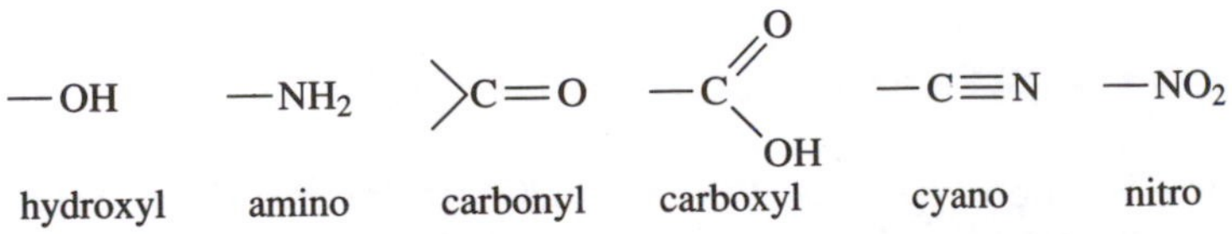

$$-\text{OH} \quad -\text{NH}_2 \quad {>}\text{C}{=}\text{O} \quad -\text{C}\overset{\displaystyle O}{\underset{\displaystyle \text{OH}}{\Big\langle}} \quad -\text{C}{\equiv}\text{N} \quad -\text{NO}_2$$

hydroxyl   amino   carbonyl   carboxyl   cyano   nitro

[1.1] EXAMPLES OF FUNCTIONAL GROUPS

The unifying principle inherent in the idea of functional groups stems from the fact that **all** the compounds that contain a particular group (for example amino, -NH_2) can be expected to have at least some chemical behaviour in common.

1.2   TYPES OF REACTION

It is all too easy to think of organic chemistry as just an enormous catalogue of the **properties** of each and every organic compound, whereas it's really about the very limited variety of things that can happen to such compounds— about their **reactions**, that is. When you first look at a textbook of organic chemistry, it seems as though there is an almost infinite number of quite different reactions that organic compounds can undergo; in fact, the number of different types of reaction is only very small:

(a) SUBSTITUTION

(b) ADDITION

(c) ELIMINATION

[1.2] TYPES OF REACTION

Substitution, addition and **elimination**—that's really all there are! So, at least in theory, perhaps things are already beginning to look a little better. Below is a simple example of each of these three different reaction types—all of them reactions with which you may well already be familiar:

(a) SUBSTITUTION: CH_3—Br + $^{\ominus}OH$ $\longrightarrow$ CH_3—OH + $Br^{\ominus}$

(b) ADDITION: CH_2=CH_2 + Br—Br $\longrightarrow$ Br—CH_2—CH_2—Br

(c) ELIMINATION: H—CH_2—CH_2—OH $\xrightarrow{H^{\oplus}}$ CH_2=CH_2 + H—OH

[1.3] TYPES OF REACTION: SIMPLE EXAMPLES

In (a) the bromine atom in CH_3—Br (bromomethane) is **substituted** by the oxygen atom of the hydroxyl ion; in (b) two bromine atoms are **added** to the molecule of CH_2=CH_2 (ethene); in (c) H and OH are **eliminated** from H—CH_2—CH_2—OH (ethanol).

1.3   BOND-BREAKING/BOND-FORMING

The three reaction types that we have just been talking about have at least one important feature in common: each of them involves the **breaking** of existing bonds between atoms (one of these atoms often being carbon), and the **forming** of new, and different, bonds.

In trying to understand what is actually happening while an organic reaction is taking place, we need to keep this bond-breaking/bond-forming idea very much in mind. In the simple substitution reaction that we looked at a moment ago, for example,

$$H_3C \ \text{---} \ Br + {}^{\ominus}OH \longrightarrow H_3C \text{---} OH + Br^{\ominus}$$

[1.4] BOND-BREAKING/BOND-FORMING: SUBSTITUTION REACTION

the really vital features are the **breaking** of the initial C—Br bond, and the **forming** of the new C—O bond.

1.4 USE OF CURLY ARROWS

As I am sure you know, a bond between two atoms involves a **pair** of electrons being shared between these atoms, which are thereby bonded to each other. We can indeed write the bond as a pair of dots (:),

$$CH_3 : Br \qquad\qquad CH_3 \text{---} Br$$

[1.5] DIFFERENT WAYS OF REPRESENTING A BOND

rather than as the more common single line (—), in order to highlight this point.

When we write a reaction down on paper, we can emphasise the vital bond-breaking/bond-forming processes by representing them through the use of **curly arrows**:

$$HO^{\ominus} : CH_3 : Br \longrightarrow HO : CH_3 + \overset{..}{Br}{}^{\ominus}$$

[1.6] USE OF CURLY ARROWS

Each curly arrow represents the movement of an electron <u>pair</u> from its original position to a new, and different, one: the **tail** of the arrow shows where the electron pair has **come from**, while the **head** of the arrow shows where the electron pair is **going to**.

In the substitution reaction in [1.6], an electron pair on the oxygen atom of the hydroxyl ion, HO:$^{\ominus}$, is moving into the space between that oxygen atom and the carbon atom of CH_3—Br to form a bond between these two atoms. At the same time the electron pair, originally shared between the carbon and bromine atoms in the C—Br bond, is moving over completely onto the bromine atom, breaking the C—Br bond, and thereby turning Br into a now detached bromide ion, Br$^{\ominus}$.

It is important to emphasise that curly arrows are an <u>exact</u> way of indicating the movement of electron pairs, and should be used with care and precision: in all too many textbooks they are sprayed about in a fashion more reminiscent of a classic Western film!

Despite any conceptual advantage of representing the electron pair in a bond by a pair of dots, convenience and ease of writing lead to use of the more common single line.

1.5   BOND POLARITY

In organic compounds the great majority of atoms, other than carbon, that are involved are—with the exception of hydrogen—more **electronegative** than carbon is:

H	C	Br	Cl	N	O	F
2.1	2.3	2.9	3.1	3.3	3.6	4.0

[1.7] RELATIVE ELECTRONEGATIVITY OF ATOMS

By more electronegative than carbon, we mean that such atoms are somewhat more successful at attracting electron pairs towards themselves than is a carbon atom. This will affect a bond, between a carbon atom and such a more electronegative atom (e.g. Br), in that the electron pair of this bond (C—Br) will not now be shared equally between the two atoms involved, but will be pulled slightly away from the carbon atom, and towards the more electronegative bromine atom: the C—Br bond will thus be **polarised** in the way indicated in [1.8].

$$C \; :Br \quad \equiv \quad C \rightarrow Br \quad \equiv \quad \overset{\delta+}{C} - \overset{\delta-}{Br}$$

[1.8] POLARITY IN A C—Br BOND

The three different representations of the polarity of the bond in [1.8] are all equally valid, and though the one representing the polarity by partial charges, i.e. $\delta+$ and $\delta-$, is probably the most graphic, it is normally more convenient just to write a simple arrow head on the bond.

1.6   TYPES OF REAGENT

1.6.1   Nucleophiles

In a bond like C—Br that is polarised in this way, the carbon atom will have had its normal quota of electrons slightly diminished (as indicated by the partial $+^{ve}$ charge in [1.8]): such a carbon atom is said to be **electron-deficient**. We would naturally expect such an electron-deficient carbon atom to be attacked preferentially by reagents that have an electron pair readily available, which they can use—by sharing it—to form a bond with the carbon atom: thereby "correcting" its electron-deficiency. Such <u>electron-rich</u> reagents are called **nucleophiles:**

$$\overset{\ominus}{:}OH \quad \overset{\ominus}{:}OEt \quad \overset{\ominus}{:}SEt \quad H_2O: \quad H_3N:$$

electron pair donors

[1.9] SOME TYPICAL NUCLEOPHILES

Nucleophiles are exactly the same kind of reagent as the more familiar **reducing agents** and **bases**: all three kinds of reagent act through providing an electron pair to share with another atom or group—all these reagents are electron pair **donors**. We have already seen ([1.3a], p. 2) a typical reaction of a carbon atom with a nucleophile in the attack of $^\ominus OH$ on

$$HO^{\ominus} \quad \overset{\delta+}{CH_3}\overset{\delta-}{-Br} \longrightarrow HO-CH_3 + Br^{\ominus}$$

[1.10] TYPICAL <u>NUCLEOPHILIC</u> SUBSTITUTION REACTION

the carbon atom in CH_3—Br. There is further discussion of this reaction in **2.1** (p. 13).

1.6.2 Electrophiles

By now you may well be thinking "if nearly all the other atoms in organic compounds are more electronegative than carbon, then should not <u>all</u> organic reactions involve attack of nucleophiles on electron-deficient carbon atoms". There are, hardly surprisingly, also situations in which a carbon atom does itself have electrons available, and is therefore **electron-rich:**

$$H_3C-CH_3 \equiv H_3C:CH_3 \qquad H_2C=CH_2 \equiv H_2C \overset{\cdot}{\underset{\cdot}{:}} CH_2$$

[1.11] ETHANE/ETHENE

As you can see, only **two** electrons are actually needed to hold together two carbon atoms—as in ethane, CH_3—CH_3, in [1.11]. It follows that in ethene, CH_2=CH_2, in which there are **four** electrons holding its two carbon atoms together (in a **double** bond), there are clearly more electrons than the two carbon atoms actually need to form a bond between them. These carbon atoms are thus electron-rich (compared with those in ethane), and this is reflected in ethene reacting very readily with reagents which are themselves

(a) $H_2C : CH_2$ $\xrightarrow[\text{H}_2\text{O}]{\text{Mn}^{\text{VII}} \rightarrow \text{Mn}^{\text{IV}}}$ $HO : CH_2 : CH_2 : OH$

(b) $H_2C : CH_2 + : \overset{..}{\underset{..}{Br}} : \overset{..}{\underset{..}{Br}} : \longrightarrow : \overset{..}{\underset{..}{Br}} : CH_2 : CH_2 : \overset{..}{\underset{..}{Br}} :$

14e   8e   +   8e

[1.12] TYPICAL <u>ELECTROPHILIC</u> ADDITION REACTIONS

deficient in electrons, and thus eager to accept them.

The two reactions shown in [1.12] have long been considered to be so characteristic of double-bonded compounds, like ethene, as to be used as diagnostic tests for them. The first reaction (a) is the permanganate oxidation of ethene, in which Mn^{VII} is accepting electrons from the ethene molecule, and being thereby converted into Mn^{IV}: in overall terms, two OH groups are added on to ethene converting it into $HO—CH_2—CH_2—OH$. There is further discussion of this reaction in (**5.1.4**, p. 77).

The second reaction (b) is the addition of bromine to ethene. The two atoms in the bromine molecule share between them a total of 14 electrons, whereas after addition to ethene to form $Br—CH_2—CH_2—Br$, these same two bromine atoms now, jointly, have a share in 16 electrons—the <u>extra</u> two electrons being supplied by, and shared with, the two carbon atoms in the addition product, $Br—CH_2CH_2—Br$. There is further discussion of this reaction in (**5.1.1**, p. 68).

Electron-deficient reagents like these are called **electrophiles** (lovers of electrons), and some typical examples are:

$$H^{\oplus} \quad {}^{\oplus}NO_2 \quad Br—Br \quad O_3 \quad AlCl_3$$

[1.13] SOME TYPICAL ELECTROPHILES

Electrophiles are exactly the same kind of reagent as the more familiar **oxidising agents** and **acids**: all three kinds of reagent act through <u>accepting</u> <u>an electron pair</u> from another atom or group, thereby becoming <u>bonded to it</u>—all these reagents are electron pair **acceptors**.

1.6.3   Radicals

It is significant that in only one of the reactions that we have seen up to now has a C—H bond been involved, and this despite the fact that this bond is by far the commonest in organic compounds. You may remember ([1.7]) that the electronegativity of hydrogen (at 2·1) is very close to that of carbon (at 2·3), which means that there will be very little polarity in a C—H bond. We would not therefore expect the carbon atom in such bonds to be attacked at all readily by either electron-rich reagents (nucleophiles), or by electron-deficient reagents (electrophiles): there is no easy way in which either type of reagent could gain a foothold at such an essentially non-polarised carbon atom.

We do indeed find in practice that a compound such as methane, CH_4, which contains **only** C—H bonds, is largely unaffected by either nucleophiles or electrophiles even under quite vigorous conditions. If, however, we just mix methane with chlorine, at room temperature, their reaction is so fast that it can even be explosive:

$$H_3C—H + Cl—Cl \longrightarrow H_3C—Cl + H—Cl$$

[1.14] REACTION OF METHANE WITH CHLORINE

You may by now be thinking "surely chlorine must be an electrophile, just like the closely similar halogen bromine was, when it added to ethene [1.12] (p. 6)?" But there is a further, interesting point about this chlorine/methane reaction—if we mix the two gases in the dark: nothing happens! If chlorine was acting as an electrophile, there seems no reason why it should not be able to do it in the dark as well. If we then shine a light on the mixture that had failed to react in the dark—off it goes like a rocket!

Now we can show that light does not have much effect on methane, but it does have a considerable effect on chlorine:

$$Cl:Cl \xrightarrow{\text{light}} Cl\cdot \quad \cdot Cl$$

[1.15] PHOTOLYSIS OF A CHLORINE MOLECULE

Light—of suitable wavelength—supplies sufficient energy to the chlorine molecule to break the bond that holds its two atoms together. The chlorine molecule is thus split into two chlorine atoms—or **radicals** as they are also called—each of them holding on to a single electron from the pair in the bond that originally joined them together: this breaking of a bond by light is called **photolysis**. Not altogether surprisingly radicals, with a single electron in their outer shell, are found to be highly reactive, and will attack the hydrogen atom of a C—H bond in CH_4 very readily. There is further discussion of this reaction in (**4.2.1**, p. 53).

1.7 EFFECT OF STRUCTURE

To date we have been considering the reactions of individual bonds in organic compounds, and we must now ask the question: to what extent can the behaviour of the bond in such a reaction be influenced by the structure of the rest of the molecule, of which this bond is a part?

We have already encountered [1.10] (p. 5) the reaction of CH_3—Br with $^{\ominus}OH$,

$$HO^{\ominus} + CH_3—Br \longrightarrow HO—CH_3 + Br^{\ominus}$$

[1.16] REACTION OF CH_3—Br WITH $^{\ominus}OH$

where the major feature of the molecule being attacked by $^{\ominus}$OH is the C—Br bond. If we were to replace one of the H atoms of the CH_3 group by a methyl group (CH_3 = Me) to form Me—CH_2—Br, what effect is this going to have on the reaction of the C—Br bond with $^{\ominus}$OH? Would we expect the reaction of Me—CH_2—Br with $^{\ominus}$OH to be faster or slower than the reaction of CH_3—Br?

What we find in the laboratory is shown in [1.17]:

$$CH_3 — Br \qquad Me — CH_2 — Br$$
$$1 \qquad\qquad 0.08$$

[1.17] RELATIVE RATES OF REACTION WITH $^{\ominus}$OH

In other words, under exactly the same conditions, bromoethane (Me—CH_2—Br) reacts with $^{\ominus}$OH about 12 times more **slowly** than bromomethane (CH_3—Br) did: if it took CH_3—Br one minute to react with $^{\ominus}$OH, it would take Me—CH_2—Br slightly more than twelve minutes.

That is an experimental observation, the question now is: how do we explain it? Well, there are really two main effects through which the rest of the molecule can influence the ease (or otherwise) with which a particular bond will react: **electronic** effects and **steric** effects.

1.7.1 Electronic effects

It is known that methyl groups (Me) are slightly **electron-donating**. Thus in bromoethane (Me—CH_2—Br) the Me group will push the electron pair, that it shares in the bond to the carbon atom of the adjacent CH_2, slightly towards that carbon atom; such electron donation is normally indicated by an arrow head (>) written on the bond (cf. [1.8], p. 4):

$$Me \rightarrow CH_2 — Br$$

[1.18] ELECTRON-DONATION BY A METHYL GROUP

You may remember that the C—Br bond is polarised by the more electronegative Br atom, so that the carbon atom of the bond carries a partial $+^{ve}$ charge, $\delta+$:

$$\overset{\delta+}{Me \rightarrow CH_2} — \overset{\delta-}{Br} \qquad\qquad \overset{\delta++}{CH_3} — \overset{\delta--}{Br}$$

[1.19] RELATIVE $+^{ve}$ POLARITY OF THE C ATOM IN Me—CH_2—Br
AND CH_3—Br

Electron-donation by the Me group in Me→CH_2—Br will decrease the $+^{ve}$ polarisation of its CH_2 carbon atom, compared with that of the CH_3 carbon atom in CH_3—Br. We would therefore expect attack by a nucleophile, such as $^{\ominus}$OH, to be somewhat more difficult on the CH_2 carbon of bromoethane than on the CH_3 carbon of bromomethane, thereby making the reaction with $Me—CH_2—Br$ the slower of the two:

$$HO^{\ominus} \quad \overset{Me}{\underset{CH_2}{\diagdown}} \overset{\delta+}{} \overset{\delta-}{—Br} \qquad\qquad HO^{\ominus} \quad \overset{\delta++}{CH_3} \overset{\delta--}{—Br}$$

[1.20] ELECTRONIC EFFECT OF Me GROUP IN
REACTION WITH $^{\ominus}$OH

Thus an electronic effect (in this case electron-donation) helps to explain the slowing down of the reaction that we observed experimentally when the Me group was introduced.

1.7.2 Steric effects

Methyl groups are considerably larger in size than hydrogen atoms, so replacement of one of the H atoms in $CH_3—Br$, by the bulkier Me group in $Me—CH_2—Br$, is likely to impede—to some extent at least—attack by $^{\ominus}$OH on the CH_2 carbon atom in $Me—CH_2—Br$, compared with similar attack on the CH_3 carbon atom in $CH_3—Br$:

$$HO^{\ominus} \quad \overset{Me}{\diagdown} CH_2—Br \qquad\qquad HO^{\ominus} \quad CH_3—Br$$

[1.21] STERIC EFFECT OF Me GROUP IN REACTION WITH $^{\ominus}$OH

Thus the steric effect of the introduced Me group will—like the electronic effect—also serve to slow down the reaction of bromoethane with $^{\ominus}$OH, compared with that of bromomethane.

Thus the electronic and steric effects of the Me group both operate in the same direction, and together provide an explanation for the reaction-slowing that we observed experimentally.

There is further discussion of the operation of both electronic and steric effects on the reaction of $^{\ominus}$OH with R—Br in (**2.1.4**, p. 18).

1.8 SUMMARY

The major problem in studying organic chemistry is how to bring some sort of order and system to the vast body of compounds, and their many reactions, that go to make up the subject. One useful generalisation is that of **functional groups**—all compounds that contain a particular group (e.g. NH_2) can be expected to have at least some chemical behaviour in common.

Analysis of the enormous range of organic reactions establishes that there are effectively only **three** different types of reaction: **substitution**, **addition** and **elimination**. The vital feature of these reactions is the **breaking** of existing bonds between atoms, and the **forming** of new, different bonds. The shifting of electron pairs from one position to another—which is what bond-breaking/bond-making entails—may be emphasised by using **curly arrows** to represent these processes.

A further helpful generalisation is that there are substantially only **three** types of reagent that attack centres (very often carbon atoms) in organic compounds: **nucleophiles** (electron-rich reagents), **electrophiles** (electron-deficient reagents) and **radicals** (reagents having an un-paired electron in their outer shell).

Finally, there are essentially only **two** effects that the rest of an organic molecule can exert on the behaviour of the particular bond that is undergoing attack: **electronic** effects and **steric** effects.

Having established these important generalisations, we will now use them to review the broad spectrum of organic reactions in a systematic way.

Substitution

2

Nucleophilic substitution

2.1 SUBSTITUTION AT A SATURATED CARBON ATOM

We have already seen a classic example of nucleophilic substitution at a saturated carbon atom in the attack of $^{\ominus}$OH on bromomethane (CH_3—Br) in [1.10] (p. 5). We have tacitly assumed, in the way we have written the reaction with curly arrows, that it proceeds *via* a simple collision between the two species involved—as common sense might well suggest:

$$HO^{\ominus} \quad CH_3{-}Br \longrightarrow HO{-}CH_3 + Br^{\ominus}$$

[2.1] NUCLEOPHILIC SUBSTITUTION OF CH_3—Br BY $^{\ominus}$OH

It would seem reasonable to suppose—in such a simple collision—that the easiest line of attack for $^{\ominus}$OH on the carbon atom would (on energy grounds) be from the side opposite to that from which the bulky Br is departing.

These suppositions prompt the question of whether such a simple pathway for this reaction is pure supposition, or whether there is some evidence to support it.

2.1.1   Kinetic evidence

We can study the **rate** at which a reaction proceeds by monitoring the **concentrations** of the species that are reacting with each other: in this case, CH_3—Br and $^{\ominus}OH$, whose concentrations will be declining as they are used up. This monitoring can be done by removing samples of the reaction mixture at regular intervals for analysis, always provided the reaction is slow enough to allow this to be done. In the case of faster reactions, the necessary monitoring can be done continuously by the use of spectroscopic, or other direct observational, methods.

The relationship between these varying concentrations and the actual rate of the reaction is called a **rate equation**, and for the reaction of CH_3—Br with $^{\ominus}OH$ it is found to have the not unexpected form:

$$RATE = k \ [CH_3 -Br] \ [\overset{\ominus}{O}H]$$

[2.2] RATE EQUATION FOR REACTION OF CH_3—Br WITH $^{\ominus}OH$

It is important to remember that the rate equation codifies the results of actual experiments done in the laboratory, and what it tells us about this particular reaction is that **both** CH_3—Br and $^{\ominus}OH$ are directly involved in controlling the rate at which it proceeds. This rate equation does not, however, **prove** that the reaction proceeds *via* a simple collision between CH_3—Br and $^{\ominus}OH$, but such a "collision pathway" is entirely compatible with a rate equation of this form.

2.1.2   S$_N$2 reaction pathway

If we do comparable experiments with $MeCH_2$—Br [1.17] (p. 8) and with Me_2CH—Br (in which another H atom of the CH_2 group has been replaced by a further Me group), we find that these reactions follow exactly similar rate equations to the one we observed with CH_3—Br:

$$RATE = k \ [R-Br] \ [\overset{\ominus}{O}H] \qquad (R = CH_3, MeCH_2 \text{ or } Me_2CH)$$

[2.3] RATE EQUATION FOR REACTION OF R—Br WITH $^{\ominus}OH$

It will be remembered (*cf.* **1.7**, p. 7) that electronic and steric effects were enlisted to account for the slowing down (≈ 12 fold) of the rate of reaction, with $^{\ominus}OH$, that was observed when one of the H atoms in the CH_3 group of CH_3—Br was replaced by an Me group. We would predict that the operation of these two effects would be enhanced when a further H atom in the CH_3 group of CH_3—Br is replaced by an Me group, i.e. in Me_2CH—Br; and that this would have the effect of slowing down the rate of reaction with $^{\ominus}OH$ still further. This is indeed what is actually observed experimentally, though the further slowing down (≈ 8 fold) of the rate of reaction produced by this second Me group is not quite as great as we might perhaps have expected.

It seems reasonable to assume that all three reactions with $^{\ominus}OH$ (of CH_3—Br, $MeCH_2$—Br and Me_2CH—Br) proceed *via* similar pathways, and—until any subsequent experimental evidence suggests otherwise—that these pathways involve similar simple collisions between the two reactants:

$$HO^{\ominus} \quad R\!-\!\mathbf{Br} \longrightarrow HO\!-\!\dot{R} + Br^{\ominus}$$

[2.4] SIMPLE COLLISION PATHWAY FOR REACTION OF
R—Br WITH $^{\ominus}OH$

Such a reaction pathway is generally described as S_N2, denoting that it refers to a **S**ubstitution reaction, is **N**ucleophilic in character, and that **2** species appear in the rate equation.

2.1.3 S_N1 reaction pathway

If we go on and replace the last of the H atoms in the original CH_3—Br by a third Me group—to form Me_3C—Br—we would expect the rate of reaction of this bromide with $^{\ominus}OH$ to be slowed down still further (compared with Me_2CH—Br). When we make the experimental measurement in the laboratory, however, the result is a great surprise:

CH_3—Br	$MeCH_2$—Br	Me_2CH—Br	Me_3C—Br
1	0.079	0.014	47.2!

[2.5] RELATIVE RATES OF REACTION OF R—Br WITH $^{\ominus}OH$

Not only is the rate of reaction of Me_3C—Br with $^{\ominus}OH$ much **faster** than that of Me_2CH—Br (≈ 360 fold), it is faster even than that of the original bromide, CH_3—Br!

There is no reason to suppose that the electronic and steric effects—which operated to slow down the reaction when first one, and then a second, H atom in CH_3—Br were replaced by Me groups—should not continue to operate when we introduce a third such group:

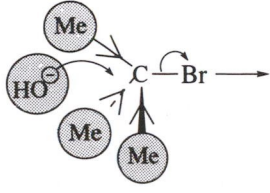

[2.6] EXPECTED OPERATION OF ELECTRONIC AND STERIC
EFFECTS IN Me_3C—Br/$^{\ominus}OH$ REACTION

The dashed and solid representations of bonds in [2.6] indicate that the atoms or groups thus bonded to carbon are <u>behind</u> and <u>in front of</u> the plane of the paper, respectively.

Clearly, therefore, some other circumstance must be at work in the speeding up of the reaction of Me_3C-Br with $^\ominus OH$ that we have not yet taken into account. When we come to do rate measurements on this reaction in the laboratory, we discover that it does not follow the same rate equation as CH_3-Br, $MeCH_2-Br$ or Me_2CH-Br did. Me_3C-Br is found to follow the rate equation:

$$\text{RATE} = k\ [Me_3C-Br]$$

[2.7] RATE EQUATION FOR $Me_3C-Br/^\ominus OH$ REACTION

This means—most surprisingly—that $^\ominus OH$ is **not** involved in controlling the rate of this reaction, which clearly cannot therefore proceed *via* a simple "collision" between $R-Br$ and $^\ominus OH$ as did the first three bromides.

This unexpected rate equation does indeed tell us that the vital breaking of the $C-Br$ bond, in the overall reaction, must involve Me_3C-Br **alone**, in some sort of "do-it-yourself" process; but how could this be achieved? If, perhaps the $C-Br$ bond in Me_3C-Br is more highly polarised than the one in CH_3-Br (or in the other two bromides), then it might be expected to undergo complete ionisation—breaking to form an **ion pair**—more readily:

$$\overset{\delta++}{Me_3C}-\overset{\delta--}{Br} \longrightarrow \overset{\delta+++}{Me_3C}\cdots\cdots\overset{\delta---}{Br} \longrightarrow \overset{\oplus}{Me_3C}\ \ \overset{\ominus}{Br}$$
$$\text{ion pair}$$

[2.8] IONISATION OF Me_3C-Br?

This immediately prompts the question of how such a change in structure could serve to promote increased polarity in, and ease of ionisation of, a $C-Br$ bond? An explanation can be offered, again based on the operation of simple electronic and steric effects.

In such ionisation of $C-Br$, the cumulative electron-donating effect of the three Me groups in Me_3C-Br will serve to stabilise—progressively—the

[2.9] BROMIDE IONISATION: ELECTRONIC EFFECT
OF 3 Me GROUPS

$+^{ve}$ charge that is developing on the carbon atom of the $C-Br$ bond, as this is undergoing ionisation: something that the three H atoms in CH_3-Br are quite unable to do.

The three Me groups in Me_3C-Br are quite large and bulky and will, in this bromide, be forced close together—crowded, in fact—in maintaining the

bromide : tetrahedral   carbocation : flat

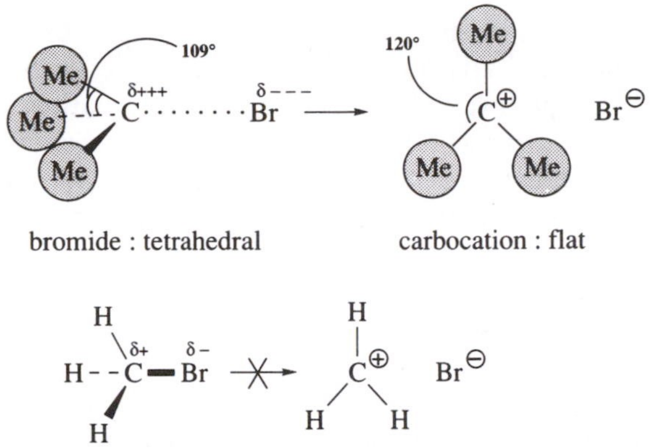

[2.10] BROMIDE IONISATION: STERIC EFFECT OF 3 Me GROUPS

tetrahedral arrangement of <u>four</u> groups about the central carbon atom, with a bond angle of $\approx 109°$. As ionisation takes place, the three Me groups will be able to move apart from each other—with consequent relief of their steric crowding—as they take up a flat (planar) arrangement in the developing **carbocation**, $Me_3C^\oplus$, which has only <u>three</u> groups attached to its central carbon atom. The bond angle in the carbocation is larger, at $\approx 120°$, and all crowding of the three bulky Me groups will now be relieved. The three H atoms in CH_3—Br are very much smaller than Me groups, and there would, therefore, be no corresponding relief of crowding if CH_3—Br were to undergo ionisation.

After such a "do-it-yourself" operation on the part of Me_3C—Br, the overall reaction could then be completed by a wholly unimpeded attack by $^\ominus OH$ on the flat (planar) $Me_3C^\oplus$ cation:

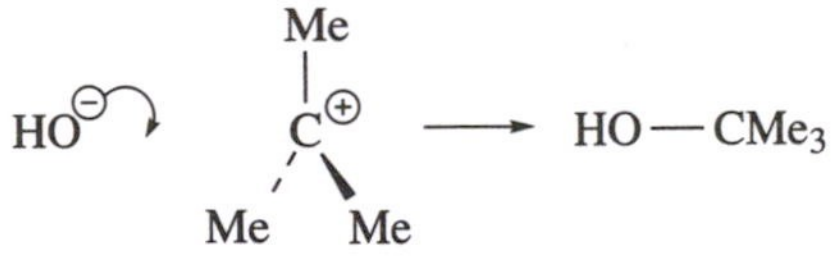

[2.11] COMPLETION OF OVERALL Me_3C—Br/$^\ominus OH$ REACTION

Such a reaction—between two ions—might well be expected to be very fast, while the preceding breaking of the C—Br bond during the ionisation of Me_3C—Br is likely to be more difficult, and hence slower. The initial step—ionisation of the bromide—would thus control the overall reaction, by setting a limit to the rate at which this could proceed. This is exactly what would be required by the rate equation, that we derived from our measurements on this reaction in the laboratory:

$$\text{RATE} = k \; [Me_3C\text{—Br}]$$

[2.12] RATE EQUATION FOR Me_3C—Br/$^\ominus OH$ REACTION

It is, however, important to emphasise that conforming to this rate equation does not **prove** that the $^{\ominus}OH/Me_3C$—Br reaction necessarily proceeds *via* such an ionisation pathway: but such a reaction pathway is entirely compatible with all experimental data that have so far been obtained.

Such a reaction pathway is generally described as S_N1, denoting that it refers to a Substitution reaction, is Nucleophilic in character, and that **1** species only appears in the rate equation. That ionisation to form carbocations really does occur, and is not merely a hypothesis devised to explain the S_N1 rate law [2.12], is demonstrated by the actual isolation of stable salts of $Me_3C^{\oplus}$ with suitable anions.

We have offered plausible, alternative reaction pathways for the nucleophilic substitution reactions of bromides, and provided an explanation—based on the operation of simple electronic and steric effects—to account for the observed shift in kinetic behaviour (rate equation). It is, however, important to emphasise that it is considerations of energy that determine which pathway is followed: the S_N1 pathway is an easier "ride" for Me_3C—Br than the S_N2 pathway would be; while, correspondingly, the S_N2 pathway is an easier "ride" for CH_3—Br than the S_N1 pathway would be.

2.1.4   Effect of structure

We can thus rationalise the effect that the structure of the rest of the molecule can have on a nucleophilic substitution reaction in quite simple terms. As the size and complexity of the alkyl groups attached to a carbon atom undergoing nucleophilic attack is increased, reaction *via* an S_N2 pathway will usually become more difficult. This is due partly to increasing obstruction to the attacking nucleophile, and partly to decreasing $+^{ve}$ polarity at the carbon atom being attacked, arising from increasing electron-donation by these groups.

On the other hand, such a change in the nature of the groups attached to the carbon atom being attacked is likely to promote a reaction *via* an S_N1 pathway. This results from the decrease in crowding, around the central carbon atom, as the arrangement about this atom changes from tetrahedral—in the original bromide—to planar—in the carbocation of the developing ion pair. The alkyl groups can also help to stabilise the developing carbocation in the ion pair, through **delocalisation** of its $+^{ve}$ charge *via* their electron-donating ability.

We have already seen this happening with the slightly electron-donating Me groups in Me_3C—Br, but the benzene ring of a **phenyl** (C_6H_5) group is considerably more effective, as reflected in the rates of reaction of the compounds below:

$$\overset{\delta++}{Me_3C}—\overset{\delta--}{Br} \longrightarrow Me_3\overset{\oplus}{C} \quad \overset{\ominus}{Br} \qquad \overset{\delta++}{(C_6H_5)_3C}—\overset{\delta--}{Br} \longrightarrow (C_6H_5)_3\overset{\oplus}{C} \quad \overset{\ominus}{Br}$$

$$1 \qquad\qquad\qquad\qquad\qquad\qquad \approx 10^8$$

[2.13] PROMOTION OF AN S_N1 PATHWAY THROUGH
CARBOCATION STABILISATION BY C_6H_5

This stabilisation of the $+^{ve}$ charge in the developing carbocation is achieved through the agency of the available electrons in the aromatic system of the phenyl groups:

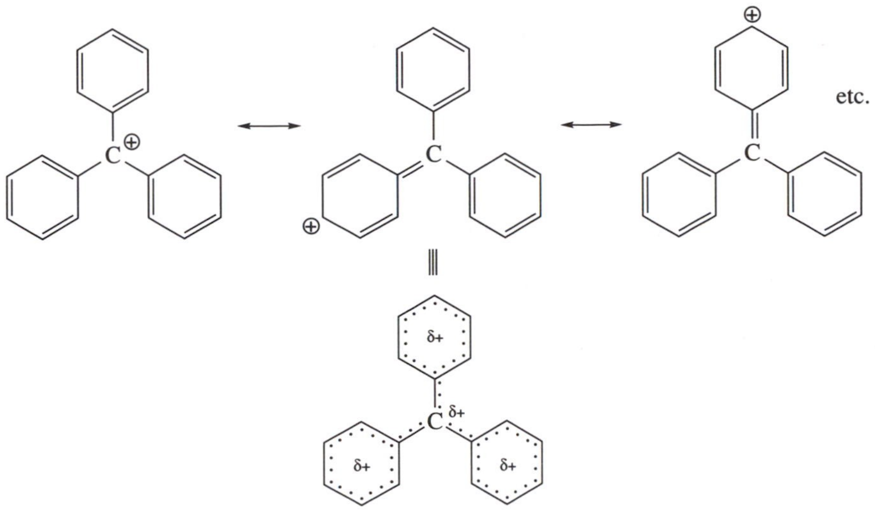

[2.14] STABILISATION OF A CARBOCATION BY PHENYL GROUPS

It is possible to write a number of alternative structures for the $(C_6H_5)_3C^\oplus$ cation, in which—through the intervention of the electrons of the three benzene rings—the $+^{ve}$ charge can be located on different carbon atoms in these rings. Three such alternative structures are represented in [2.14], and the fact that they differ from each other only in the location of the electron pairs is commonly indicated by writing a double-headed arrow, $\leftrightarrow$, between them. Such alternative representations are referred to as **canonical structures**, and no single one of them is, by itself, an adequate representation of the $Ph_3C^\oplus$ cation. The best representation of the "real" structure of the cation, at least in terms of electron distribution, is the one shown on the lower line in [2.14] in which the $+^{ve}$ charge is delocalised over the three benzene rings, and the cation thereby stabilised. The stabilisation is indeed such that Ph_3C—Br is found to be largely converted into $Ph_3C^\oplus Br^\ominus$ in solution in liquid SO_2. The question of delocalisation of $+^{ve}$ charge in aromatic systems is considered further in **3.2.1.2** (p. 34), [3.27] (p. 43), and [3.28] (p. 45).

Similar promotion of reaction occurs with the electron system of a simple double bond attached to a carbon atom undergoing nucleophilic attack, but somewhat less effectively:

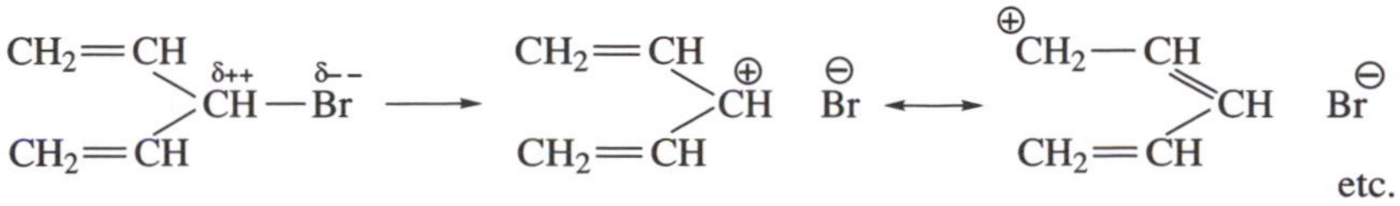

[2.15] STABILISATION OF A CARBOCATION BY DOUBLE BONDS

2.1.5   Effect of solvent

Changing the solvent in which a particular nucleophilic substitution reaction is carried out is found to have a tremendous effect on the rate at which it takes place. Thus, in as simple a reaction as you can readily think of—in which a chloride ion attacks iodomethane—

$$Cl^{\ominus} \quad \overset{\delta+}{CH_3}\overset{\delta-}{-I} \longrightarrow Cl-CH_3 + I^{\ominus}$$

MeOH: 1
HCONMe$_2$: 1.2×10^6

[2.16] EFFECT OF SOLVENT ON CH$_3$—I/Cl$^{\ominus}$ REACTION

changing the solvent from methanol (MeOH) to *N,N*-dimethylformamide (HCONMe$_2$) is found to result in a **million-fold** (10^6) increase in the rate of reaction!

The reason for this amazing difference is believed to be that in MeOH the negatively charged nucleophile, Cl$^{\ominus}$, is surrounded by molecules of the solvent methanol, which are attached to it through **hydrogen bonds**:

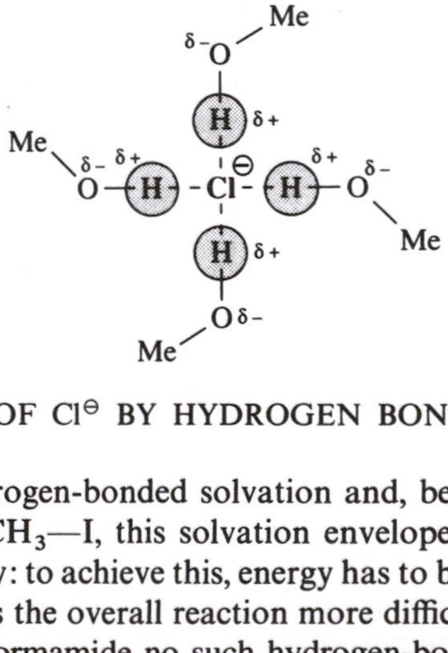

[2.17] SOLVATION OF Cl$^{\ominus}$ BY HYDROGEN BONDING WITH MeOH

This is known as hydrogen-bonded solvation and, before the Cl$^{\ominus}$ ion is in a position to attack CH$_3$—I, this solvation envelope of MeOH molecules has to be stripped away: to achieve this, energy has to be expended. This need for energy input makes the overall reaction more difficult, and hence slower.

With *N,N*-dimethylformamide no such hydrogen-bonded solvation of Cl$^{\ominus}$ is possible, as the ability to form hydrogen bonds is essentially confined to hydrogen attached to highly electronegative atoms such as O and N. There is thus no solvation envelope to strip away from Cl$^{\ominus}$, the energy demand is smaller, the reaction is easier, and hence faster. What we are saying is that an unsolvated Cl$^{\ominus}$ ion is a very much more powerful nucleophile than the same ion hidden inside a solvation envelope.

That it is the effectiveness of hydrogen-bonded solvation which is responsible for the slowing of the reaction in MeOH is borne out by the behaviour of the same reaction in *N*-methylformamide (HCONHMe):

$$Cl^{\ominus} \quad \overset{\delta+}{CH_3} \overset{\delta-}{-I} \longrightarrow Cl-CH_3 + I^{\ominus}$$

MeOH: 1
HCONHMe: 45.3

[2.18] EFFECT OF SOLVENT ON $Cl^{\ominus}/CH_3$—I REACTION

Despite this solvent being very similar in structure to dimethylformamide, and quite unlike methanol, the rate of reaction is quite similar to that in methanol, but getting on for a hundred thousand times slower than in the structurally very similar $HCONMe_2$. The reason is, of course, that N-methylformamide still contains an N—H linkage that is capable of hydrogen-bonding with $Cl^{\ominus}$, and thus of slowing down the rate of its reaction with CH_3—I; though not quite so effectively as did MeOH, which can form stronger hydrogen bonds.

The above reaction proceeds *via* the S_N2 pathway—like the very similar CH_3—Br/$^{\ominus}$OH reaction that we have already considered. For a nucleophilic substitution that proceeds *via* the S_N1 pathway, the reaction will be speeded

$$\overset{\delta++}{Me_3C} - \overset{\delta--}{Br} \xrightarrow[\text{slow}]{} \overset{\oplus}{Me_3C} \quad \overset{\ominus}{Br}$$

EtOH: 1
50% H_2O/50% EtOH: $\approx$ 30,000

[2.19] S_N1 PATHWAY: RATE-LIMITING STEP

up as the solvent becomes more polar (as measured by its **dielectric constant**). This is because the more polar the solvent, the less the energy that has to be supplied to effect the separation of the two ions in the ion pair developing during the rate limiting step of the reaction.

Thus in the reaction of Me_3C—Br with $^{\ominus}$OH, changing the solvent from CH_3CH_2OH (ethanol) to the more polar 50% H_2O/50% CH_3CH_2OH increases the rate of reaction 30,000 fold! Hydroxylic solvents are particularly effective in this respect because they are able to stabilise—through use of their H and O atoms—**both** of the developing ions in the ion pair, as the latter is forming:

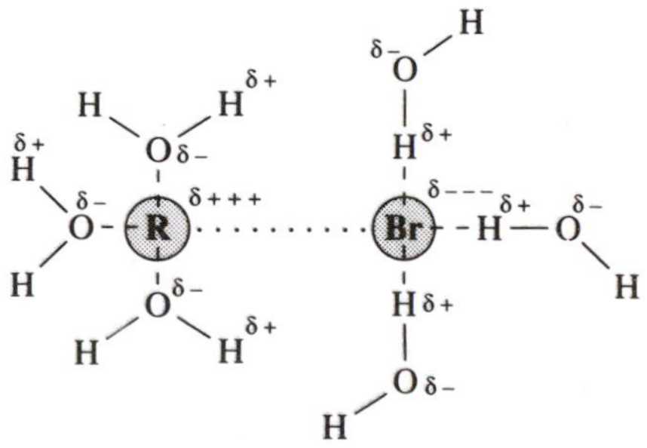

[2.20] STABILISATION OF BOTH IONS IN A DEVELOPING ION PAIR

The importance of such solvation in S_N1 reactions is borne out by the fact that reactions proceeding *via* this pathway are very uncommon in the gas phase.

2.1.6   Effect of entering group

We have to date considered only $^\ominus$OH and Cl$^\ominus$ as potential nucleophiles, although we have seen some others listed in **1.6.1** (p. 5). As was stated then, it is not necessary for a nucleophile to be negatively charged; what a nucleophilic species has to be capable of doing is sharing an electron pair with the atom being attacked, which is often carbon.

One might, therefore, expect there to be some relationship between the effectiveness of a species as a nucleophile and its strength as a base, because both activities involve the donation of an electron pair. It would be very helpful if there were such a relationship, because data on basic strength is readily available for a wide range of species, while reliable data on nucleophilic effectiveness is not all that easy to come by.

Acting as a base commonly involves donation of an electron pair to hydrogen, while acting as a nucleophile involves similar donation to other atoms—often to carbon. Despite this difference in the atom to which the electron pair is donated, we find in practice that the two abilities quite often do run in parallel. This is so in the series in [2.21], provided that in all the species being compared the atom involved in nucleophilic, or basic, activity is the same—in this series, an <u>oxygen</u> atom:

$$EtO^\ominus > C_6H_5O^\ominus > MeCO_2^\ominus > NO_3^\ominus$$

[2.21] BASIC STRENGTH/NUCLEOPHILIC ABILITY RUNNING
IN PARALLEL

If, however, we compare nucleophile/base species whose electron-donating atom is different, then the two abilities often do <u>not</u> run in parallel. This is the case in [2.22], where the two abilities are found to run in precisely

basic strength: $EtO^\ominus > EtS^\ominus$ $F^\ominus > Cl^\ominus > Br^\ominus > I^\ominus$

nucleophilic ability: $EtS^\ominus > EtO^\ominus$ $I^\ominus > Br^\ominus > Cl^\ominus > F^\ominus$

[2.22] BASIC STRENGTH/NUCLEOPHILIC ABILITY <u>NOT</u> RUNNING
IN PARALLEL

opposite directions! This divergence of nucleophilic ability from basic strength stems from the fact that as the atom donating the electron pair increases in **size**, the electrons in its outer shell will be further away from—and hence held less tightly by—the atomic nucleus. These outer electrons are thus more **polarisable**: they are more readily available to form a bond with the atom being attacked. Polarisability appears to be much more important in nucleophilic ability than in the equilibrium situation involved in basicity; thus species in which the relevant atom is large are commonly found to be better nucleophiles than their strength as bases might suggest.

As was stated earlier in this section, it is not necessary for a species to carry a negative charge before it can act as a nucleophile, and many excellent nucleophiles are uncharged. A case in point is a nitrogen atom, with an electron pair available:

$$H_3N\colon\ \ R\!-\!Br \longrightarrow H_3\overset{\oplus}{N}\!-\!R\ \ \overset{\ominus}{Br}$$

[2.23] NUCLEOPHILIC NITROGEN

There is also a large variety of entering groups, both charged and uncharged, in which the nucleophilic atom is sulphur, oxygen, or even carbon.

Finally, the nucleophilic ability of a species, which is obviously of great importance in S_N2 reactions (cf. **2.1.2**, p. 14), will be of little significance in S_N1 reactions because the attacking nucleophile then plays no part in controlling the rate of the overall reaction (**2.1.3**, p. 15).

2.1.7 Effect of leaving group

In a nucleophilic substitution reaction such as that in [2.24], the leaving group, Y, will clearly influence reaction *via* either S_N1 or S_N2 pathways,

$$\overset{\ominus}{HO}+\ R\!-\!Y \longrightarrow HO\!-\!R\ +\ \overset{\ominus}{Y}$$

[2.24] LEAVING GROUP IN NUCLEOPHILIC SUBSTITUTION

because breaking the R—Y bond is directly involved in controlling the rate of reaction *via* either pathway (**2.1.2**, p. 14; **2.1.3**, p. 15). We might thus expect the **strength** of the C—Y bond in R—Y to play a part, as it is seen to do in the sequence in [2.25]:

$$R\!-\!I\quad R\!-\!Br\quad R\!-\!Cl\quad R\!-\!F$$

bond strength: $\longrightarrow$

leaving group ability: $\longleftarrow$

[2.25] LEAVING GROUP ABILITY AND BOND STRENGTH

The stronger the C—Y bond, the poorer Y is as a leaving group, and thus the slower the reaction of R—Y is with a nucleophile.

This is one of the reasons why nucleophiles are unlikely to react with simple C—H bonds, which are usually very strong. The case of C—H illustrates another important requirement for a leaving group in R—Y: namely increasing stabilisation of Y—through solvation or otherwise—as it is being converted into $Y^\ominus$. Thus in the hypothetical reaction of R—H with $Br^\ominus$ as nucleophile,

$$Br^\ominus + R—H \longleftarrow \overset{\frown}{Br}—R \overset{\ominus}{H}$$

[2.26] "REACTION" OF $Br^\ominus$ WITH R—H

the potential leaving group hydride ion, $H^\ominus$, far from being stabilised, is in fact extremely reactive. It is indeed a very powerful nucleophile, such that reaction does take place very readily: but in the direction opposite to the one we were expecting! $H^\ominus$ is also an extremely powerful reducing agent (cf. **7.2.4**, p. 111).

As is apparent from the several examples we have now seen, halide ions are very good leaving groups—especially the iodide ion, $I^\ominus$. The anions of strong acids tend to be good leaving groups—for example, $CF_3SO_3^\ominus$, which is excellent—because they are very stable as anions, and can be stabilised still further through solvation, especially in hydroxylic solvents.

2.2   SUBSTITUTION AT AN UNSATURATED CARBON ATOM

We find that when, for example, a chlorine atom is attached to an unsaturated carbon atom in such compounds as chloroethene (vinyl chloride, CH_2=CHCl), and chlorobenzene (C_6H_5Cl), attack by nucleophiles (e.g. $^\ominus OH$) is often extremely difficult. Thus the reaction of chlorobenzene (C_6H_5Cl) with $^\ominus OH$ requires temperatures in excess of 200°C, compared with simple alkyl halides which react readily at room temperature, or not much above.

$$
\begin{array}{c}
Cl \\
| \\
CH \\
\| \\
CH_2
\end{array}
$$

[2.27] CHLORINE ATTACHED TO UNSATURATED CARBON ATOMS

There are a number of reasons for this lack of reactivity in such halides. Thus the increased electron density associated with unsaturated carbon atoms [1.11] (p. 5) will serve to discourage the approach of a nucleophile in an S_N2 reaction pathway. The electron density at this carbon atom will also be increased because such an unsaturated carbon is able to draw the electron pair of the C—Cl bond more towards itself, than is a saturated carbon atom. This also has the effect of strengthening the C—Cl bond, and of making loss of chlorine as $Cl^\ominus$ more difficult, thus discouraging the ionisation required by an S_N1 pathway.

2.2.1 S_N2 (aromatic) reaction pathway

While attack by $^{\ominus}OH$ on chlorobenzene (C_6H_5Cl) itself is difficult, introduction into the benzene ring of electron-withdrawing substituents, such as nitro groups (NO_2), is found to raise the rate of reaction considerably:

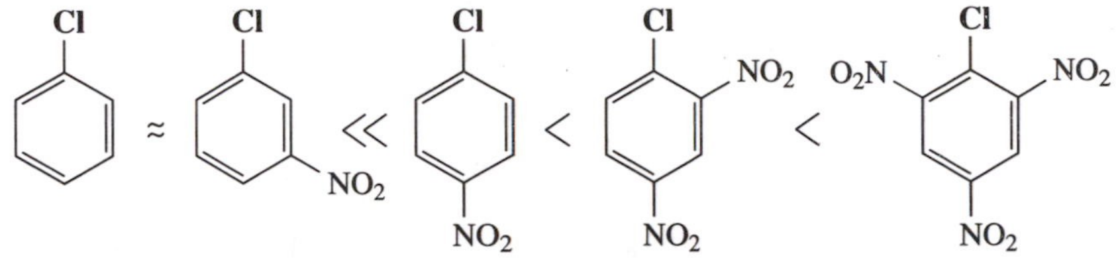

[2.28] EFFECT OF NO_2 SUBSTITUENTS ON RATE OF REACTION OF C_6H_5Cl WITH $^{\ominus}OH$

We find that these, and many similar reactions, follow the rate equation in [2.29], and while this exactly parallels the one we found for S_N2 reactions

$$RATE = k\,[Ar–Cl]\,[Nu^{\ominus}]$$

[2.29] RATE EQUATION FOR Ar—Cl/NUCLEOPHILE REACTIONS

of alkyl bromides (**2.1.2**, p. 14) there are differences. Thus in S_N2 reactions of alkyl halides, R—Hal, the nature of the leaving group **does** influence the rate of the overall reaction, e.g. R—I reacts faster than R—Cl; it **does not** normally influence the rate of the overall reaction of aryl halides, Ar—Hal. This establishes that while breaking the C—Hal bond in alkyl halides **is** involved in control of the overall reaction rate, the breaking of the C—Hal bond in aryl halides **is not**.

We have implicitly interpreted (in [2.1], p. 13) the S_N2 pathway for alkyl halides in terms of a smooth transition from reactants to products, in which forming of the bond to the nucleophile, and breaking of the bond to the leaving group, go on simultaneously in a single step:

[2.30] SIMULTANEOUS BOND-FORMING/BOND-BREAKING IN S_N2 PATHWAY FOR R—Hal

The reaction of Ar—Hal with $^{\ominus}OH$ clearly cannot occur *via* such a single step concerted pathway, as breaking of the C—Hal bond is not involved in controlling the overall reaction rate. The step controlling the rate of the overall reaction must involve both Ar—Hal and $^{\ominus}OH$, as both appear in the rate equation in [2.29] above, but in a way that does not involve breaking the C—Hal bond in Ar—Hal.

The simplest interpretation is the slow, rate-limiting formation of an **intermediate**, from which the leaving group, in this case Cl$^\ominus$, is lost in a subsequent rapid step, which does not influence the rate of the overall reaction. Such a reaction pathway is generally referred to as S_N2(**aromatic**), and does actually involve addition/elimination.

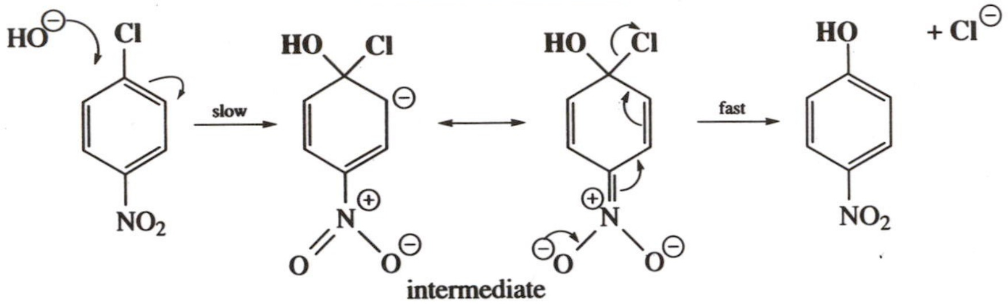

[2.31] FORMATION OF AN INTERMEDIATE IN THE
S_N2 (aromatic) PATHWAY

The role of groups such as NO_2, in promoting reaction by such a pathway, stems in part from their being able, through electron-withdrawal, to increase the $+^{ve}$ character of the carbon atom of the C—Cl bond, thereby making attack on it by $^\ominus$OH easier. More significant, however, is the ability of electron-withdrawing groups to stabilise, through delocalisation, the $-^{ve}$ charge developing on the S_N2 (aromatic) intermediate, thereby making its formation easier.

The question then arises of whether there is any satisfactory evidence for the existence of such species as the anionic intermediate in [2.31] above? In this connection, it has proved possible to isolate the same red potassium salt, [2,4,6-(NO$_2$)$_3$,1-EtOC$_6$H$_2$OMe]$^\ominus$K$^\oplus$, **either** from the reaction of

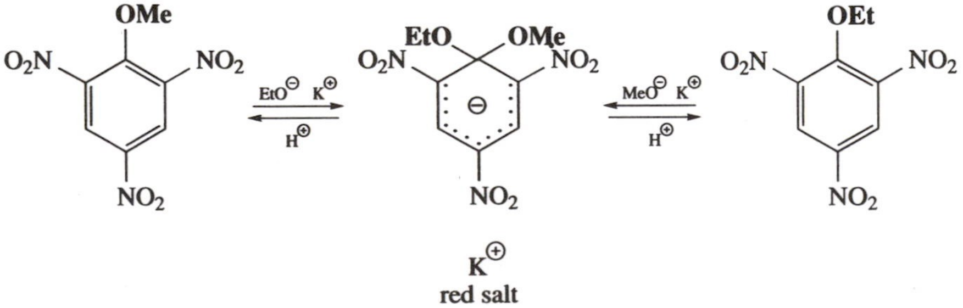

[2.32] FORMATION OF AN S_N2 (aromatic) INTERMEDIATE

EtO$^\ominus$K$^\oplus$ with the MeO compound in [2.32] **or** of MeO$^\ominus$K$^\oplus$ with the EtO compound, and to establish its structure by spectroscopic, and X-ray, methods. While it has not proved possible to isolate similar intermediates, during the course of nucleophilic attack by $^\ominus$OH on aromatic chloro compounds (probably because $Cl^{\ominus}$ is such a good leaving group), the very existence of the red intermediate in [2.32] does make the involvement of similar intermediates, in S_N2 (aromatic) reactions in general, very much more plausible.

2.2.2 Aryne reaction pathway

We have already referred to the lack of reactivity of chlorobenzene (C_6H_5Cl) towards $^{\ominus}OH$, and to how a temperature of $200\,°C$ is required before reaction will take place. However, on seeking to react chlorobenzene with a different nucleophile, amide ion ($^{\ominus}NH_2$), reaction is found to take place

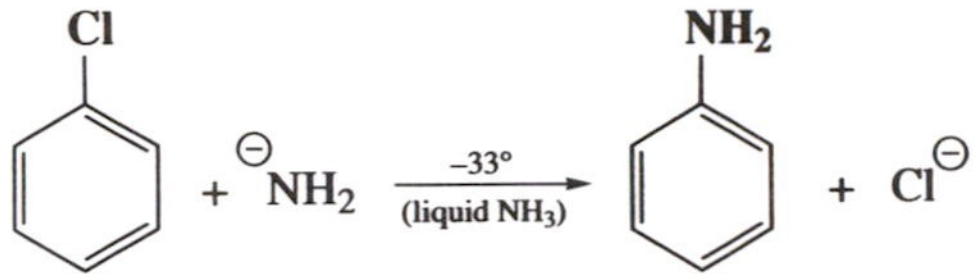

[2.33] REACTION OF C_6H_5Cl WITH AMIDE ION ($^{\ominus}NH_2$)

quite rapidly even at $-33\,°C$ (in liquid ammonia)! This great difference in ease of reaction of chlorobenzene, with two nucleophiles as similar as $^{\ominus}OH$ and $^{\ominus}NH_2$, suggests that $^{\ominus}NH_2$ may perhaps be reacting *via* a route other than the S_N2 (aromatic) pathway.

The first clue to what could be happening is provided by what is observed when $^{\ominus}NH_2$ reacts with $p\text{-MeC}_6H_4Cl$:

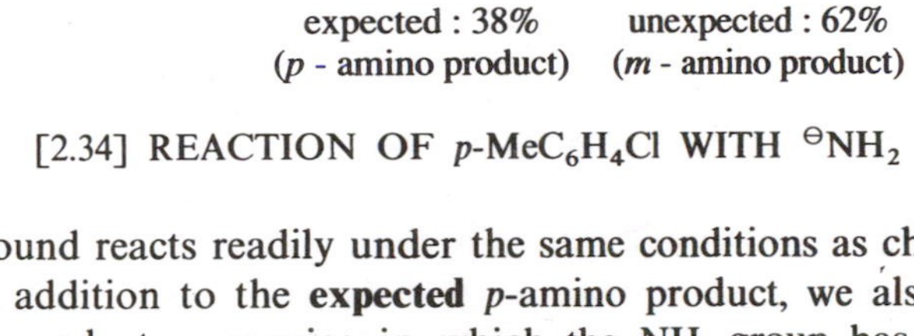

expected : 38%   unexpected : 62%
(*p* - amino product)   (*m* - amino product)

[2.34] REACTION OF $p\text{-MeC}_6H_4Cl$ WITH $^{\ominus}NH_2$

This compound reacts readily under the same conditions as chlorobenzene did but, in addition to the **expected** *p*-amino product, we also obtain an **unexpected** product: a species in which the NH_2 group has entered the benzene ring in a position <u>adjacent</u> to the one from which the chlorine atom has been lost, i.e. Cl <u>cannot have</u> been displaced <u>directly</u> by the entering NH_2 group.

Any suggestion that this <u>unexpected</u> product is formed through transformation of the first-formed, expected, product, during the course of the reaction, can be ruled out by separate experiments which establish that neither product can be converted into the other under the conditions of the reaction. It is found that in the reaction of p-MeC$_6$H$_4$Cl with $^\ominus$NH$_2$ **only** m- and p-amino products are formed, **never** the o-amino product.

The most obvious difference in properties between $^\ominus$OH and $^\ominus$NH$_2$ is that the latter is a very much stronger **base** (in liquid NH$_3$) than the former (in H$_2$O); so much so that $^\ominus$NH$_2$ is known to be able to remove H, as a proton, from a benzene ring ($^\ominus$OH cannot do so). This suggests that the reaction of $^\ominus$NH$_2$ with p-MeC$_6$H$_4$Cl could perhaps be initiated by proton removal, rather than by the expected attack on the C—Cl carbon atom:

[2.35] ARYNE PATHWAY FOR p-MeC$_6$H$_4$Cl/$^\ominus$NH$_2$ REACTION

Such a loss of a proton would be promoted by the electron-withdrawing Cl atom on the adjacent carbon: which is why this proton is removed rather than one of those o- to the Me group. Proton removal could be followed by the loss of the good leaving group Cl$^\ominus$, from the adjacent carbon atom, to form an **aryne** intermediate. We would expect this intermediate (here written with something approaching a triple bond, though molecular geometry would rule out anything resembling the situation in ethyne, HC≡CH) to be highly reactive, and to undergo ready overall addition of NH$_3$, NH$_2$ becoming bonded to one carbon atom and H to the other. This addition of NH$_3$ could take place in either of two different ways round, to yield the expected, and unexpected, products, respectively.

We would not expect these two products to be obtained in equal amounts, as the two positions available for attack in the aryne intermediate are not equivalent to each other—their orientation with respect to the Me substituent is different. It should perhaps be emphasised that though the aryne pathway results, overall, in substitution it does actually involve elimination/addition!

As usual, this prompts the question of whether there is any independent evidence for such a reaction pathway and, in particular, for the existence of arynes themselves. We find that the Cl compound in [2.36], with one o-Me substituent, reacts as readily with $^\ominus$NH$_2$ in liquid NH$_3$ as did C$_6$H$_5$Cl itself;

and the reaction leads to formation of the two amino compounds that we would now expect:

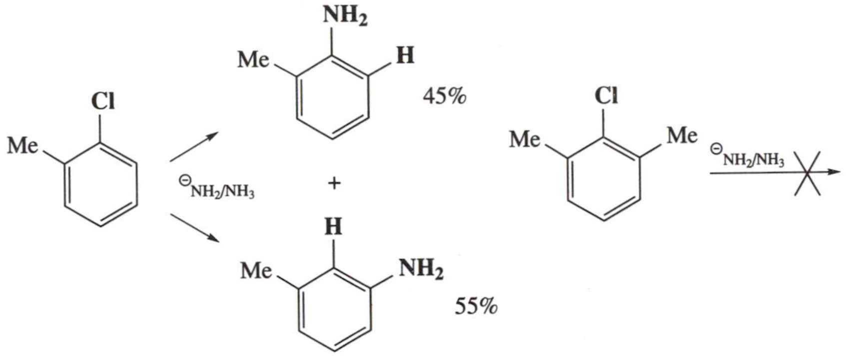

[2.36] REACTION OF o-MeC$_6$H$_4$Cl AND 2,6-Me$_2$C$_6$H$_3$Cl WITH $^\ominus$NH$_2$

By contrast, we find that the Cl compound in [2.36], with Me substituents in **both** o-positions (and hence no o-H atoms that could be removed by $^\ominus$NH$_2$ to initiate aryne formation), is entirely unaffected by $^\ominus$NH$_2$ in liquid NH$_3$, and requires conditions resembling those employed for C$_6$H$_5$Cl/$^\ominus$OH before reaction will take place. The tremendous change in reactivity, that occurs on introducing the second Me group, seems much too great to be due to the operation of any possible steric effect. It is, however, entirely compatible with the aryne pathway now being blocked, and a consequent shift to the more demanding S$_N$2 (**aromatic**) pathway (**2.2.1**, p. 25) being required if any reaction is to take place.

It has been possible to detect benzyne (C$_6$H$_4$) spectroscopically at very low temperature, and numerous methods have been developed of generating arynes in solution in suitable solvents—so much so that they are now crucial to a number of regular synthetic procedures.

2.3 SUMMARY

Substitution at carbon atoms by **nucleophiles**—electron-rich reagents—is promoted by electron-withdrawing groups attached to the carbon atom being attacked. Kinetic evidence suggests that nucleophilic substitution at a **saturated** carbon atom proceeds by one or other of two different pathways: (1) a simple one-step collision between the nucleophile and the molecule being attacked; this is known as the S$_N$2 pathway—Substitution Nucleophilic in which **2** species are involved in the kinetic rate equation; (2) slow, "do-it-yourself" loss of the leaving group from the carbon atom being attacked; this is known as the S$_N$1 pathway, as only **1** species is involved in the kinetic rate equation; it is followed by rapid, non rate-limiting attack of the nucleophile in a second step.

Consideration is then given to the influence of structure (in the molecule being attacked), of the solvent, of the leaving group, and of the entering group (the nucleophile) on the course of these two pathways.

Attention is then drawn to the lower reactivity of an **unsaturated** carbon atom towards nucleophilic substitution; in particular to nucleophilic attack on an **aromatic** carbon atom (one in a benzene ring) that carries a potential leaving group—the S_N2 (**aromatic**) pathway, and to how this differs from the S_N2 pathway for attack at a saturated carbon atom. Finally, mention is made of the alternative pathway for attack at an aromatic carbon atom, where the attacking nucleophile is also a very strong base, which involves **aryne** intermediates.

3

Electrophilic substitution

3.1 SUBSTITUTION AT A SATURATED CARBON ATOM

In theory at least there could be electrophilic substitution at a saturated carbon atom by reaction pathways essentially analogous to those we observed for nucleophilic attack:

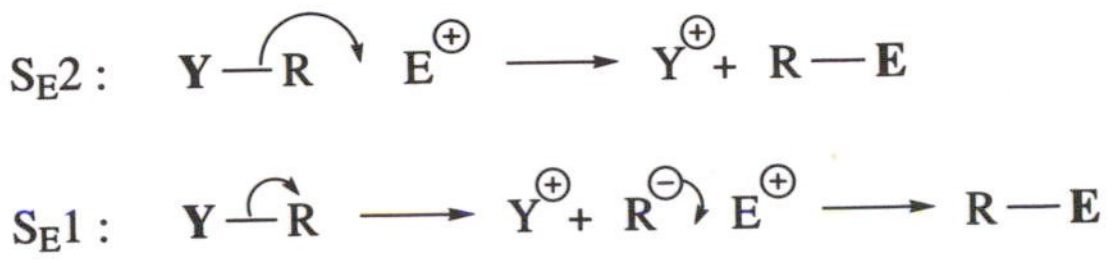

$$S_E2: \quad Y-R \quad E^\oplus \longrightarrow Y^\oplus + R-E$$

$$S_E1: \quad Y-R \longrightarrow Y^\oplus + R^\ominus \, E^\oplus \longrightarrow R-E$$

[3.1] PATHWAYS FOR ELECTROPHILIC SUBSTITUTION AT A
SATURATED CARBON ATOM

This time the single step, concerted pathway would be designated S_E2, where E now indicates that the reaction at carbon involves attack by an electrophile; while the two-step pathway involving ionisation would be designated S_E1.

For electrophilic attack on R—Y, the leaving group Y has to depart leaving the electron pair of the original C—Y bond (in R—Y) behind on R. As we have already seen (**1.5**, p. 4), nearly all the Y atoms that might be present in a C—Y bond are more electronegative than carbon, and are thus more likely to depart with the electron pair than is carbon. We should thus not expect electrophilic attack to occur at a saturated carbon atom unless the Y atom of the C—Y bond was rather less electronegative than carbon; only then would the carbon atom be able to draw the electron pair (of the C—Y bond) away from Y and towards itself.

One class of Y atoms that meets this criterion are metals, and a range of organometallic compounds of saturated carbon are indeed found to undergo electrophilic attack:

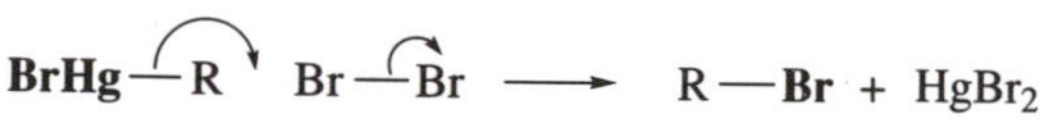

$$\text{BrHg} \overgroup{\text{—R}} \quad \text{Br} \overgroup{\text{—Br}} \longrightarrow \quad \text{R—Br} + \text{HgBr}_2$$

[3.2] ELECTROPHILIC ATTACK AT SATURATED CARBON IN AN ORGANOMETALLIC COMPOUND

3.2   SUBSTITUTION AT AN UNSATURATED (AROMATIC) CARBON ATOM

Electrophilic attack is of much greater importance at an unsaturated carbon atom, though we might well have expected this to result in addition at the electron-rich carbon atoms (**1.6.2**, p. 6) rather than in substitution. When, however, the unsaturated carbon atoms are part of an aromatic system then substitution is indeed found to occur. Among the best known examples of such a substitution reaction is **nitration**.

3.2.1   Nitration

When benzene, C_6H_6, is treated with a mixture of concentrated nitric and sulphuric acids, a hydrogen atom attached to the ring is substituted by a nitro, NO_2, group:

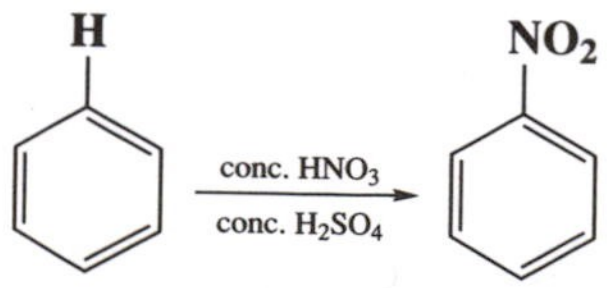

[3.3] NITRATION OF BENZENE

The first point of interest about this reaction is that it proceeds very slowly—if at all—with concentrated HNO_3 alone; concentrated H_2SO_4 is also required. Under the conditions of the reaction concentrated H_2SO_4, by itself, is found to have no effect on benzene, so its vital role in the overall reaction must be in connection with the concentrated HNO_3.

3.2.1.1   The nature of the electrophile

If we measure the freezing point of a solution of concentrated HNO_3 in concentrated H_2SO_4, we find that this is lowered compared with that of concentrated H_2SO_4 itself. This lowering of the freezing point is, however, found to be approximately **four times** as great as would, in theory, have been expected from the amount of concentrated HNO_3 that had been dissolved in the sulphuric acid. This indicates that every molecule of HNO_3, originally dissolved in the concentrated H_2SO_4, has been converted, in the acid mixture, into **four** new species.

Sulphuric is a stronger acid than nitric, and we can envisage it protonating the weaker HNO_3:

$$O_2N{-}OH + H_2SO_4 \rightleftharpoons O_2N{-}\overset{\oplus}{\underset{H}{OH}} + HSO_4^{\ominus} \rightleftharpoons \overset{\oplus}{NO_2} + H_2O + HSO_4^{\ominus}$$

$$\Big\updownarrow H_2SO_4$$

$$H_3\overset{\oplus}{O} + HSO_4^{\ominus}$$

i.e. $$HNO_3 + 2H_2SO_4 \rightleftharpoons \overset{\oplus}{NO_2} + H_3\overset{\oplus}{O} + 2HSO_4^{\ominus}$$

[3.4] FORMATION OF NITRONIUM ION ($^{\oplus}NO_2$)

The protonated HNO_3 contains a very good potential leaving group, in H_2O, and loss of this would result in the formation of $^{\oplus}NO_2$, a **nitronium ion**; the H_2O that is also formed would then, in turn, undergo protonation by the sulphuric acid. The overall result is the formation of <u>four</u> species from each molecule of HNO_3 originally dissolved: $^{\oplus}NO_2$, $H_3O^{\oplus}$ and $2HSO_4^{\ominus}$. The most significant point is the formation—in $^{\oplus}NO_2$—of a powerful potential electrophile, whose presence in the solution can indeed be detected spectroscopically. The role of $^{\oplus}NO_2$ as the effective electrophile is strongly supported by the observation that very easy nitration of benzene, and of other aromatic species, can be effected by compounds known to contain $^{\oplus}NO_2$, such as the salt, $^{\oplus}NO_2 \, BF_4^{\ominus}$.

3.2.1.2   Kinetics and the reaction pathway

Many nitration reactions of aromatic species are found to follow the slightly idealised rate equation (Ar is used to indicate an aromatic group):

$$\text{RATE} = k \ [\text{Ar}-\text{H}] \ [\overset{\oplus}{\text{NO}_2}]$$

[3.5] IDEALISED RATE EQUATION FOR NITRATION

With a rate equation such as this indicating that <u>both</u> Ar—H and $^{\oplus}\text{NO}_2$ are involved in controlling the overall reaction, nitration could proceed by any one of several different pathways:

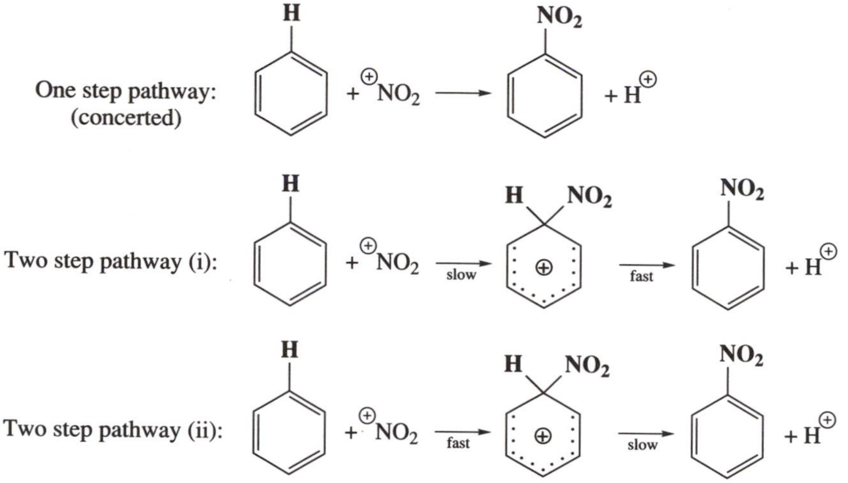

One step pathway: (concerted)

Two step pathway (i):

Two step pathway (ii):

[3.6] ONE STEP (CONCERTED) OR TWO STEP (*VIA* AN INTERMEDIATE) PATHWAYS?

Thus it could follow a concerted pathway (*cf.* S_E2 in [3.1], p. 31), in which the C—H bond is being broken and the C—NO_2 bond being formed, simultaneously, in a single step; or it could follow alternative two step pathways, involving an <u>intermediate</u> species, in which <u>either</u> of the two steps could be the slower one that controls the overall rate of the reaction. In the two pathways that proceed *via* an intermediate, this has been written with the $+^{\text{ve}}$ charge—brought by $^{\oplus}\text{NO}_2$—delocalised over the unsaturated system of the benzene ring. Such delocalisation serves to stabilise the cationic intermediate, and thereby make its formation easier.

3.2.1.3   Deciding between different pathways

The question then arises of whether it is possible to distinguish between these different possible pathways, and thus to determine which, if any, of them is likely to be followed. An answer can be provided that depends on the fact that the bond between carbon and the heavier isotope of hydrogen deuterium (D), a C—D bond, is stronger than the corresponding bond between carbon and hydrogen, C—H. It is found in the laboratory that when exactly similarly situated C—D and C—H bonds undergo the same reaction under identical conditions, then the reaction of the C—D bond is approximately seven times slower at room temperature: this is known as a **kinetic isotope effect**.

If we nitrate benzene (C_6H_6) and hexadeuterobenzene (C_6D_6) under identical conditions, and compare the rates of the two reactions, what we

$\dfrac{k_H}{k_D} \approx 1.00$ (at 25°C)

[3.7] NITRATION OF C_6H_6 AND C_6D_6

actually find is that both reactions proceed at exactly the same rate! What this tells us is, of course, that in the nitration of benzene (and of hexadeuterobenzene) the breaking of the C—H (or C—D) bond **cannot** be involved in the step that controls the overall rate of the nitration reaction: if it were the two reactions would necessarily proceed at different rates.

This immediately rules out the one step concerted pathway in [3.6] (p. 34), because the C—H bond **must** be broken during the course of the single step by which it proceeds. Further, of the alternative two step pathways—(i) and (ii) in [3.6] (p. 34)—(ii) must also be ruled out because in this pathway the breaking of the C—H bond is the slow step, which would thus control the rate of the overall reaction.

This leaves two step pathway (i), in which initial attack by $^\oplus NO_2$ is the slow step which controls the overall rate of nitration, and this is indeed compatible with the observed absence of any kinetic isotope effect. It is important to emphasise that these considerations do not, however, **prove** that the nitration of benzene proceeds *via* two step pathway (i); but we can say that such a pathway is in agreement with all the experimental evidence to date.

3.2.1.4 Substitution versus addition

It is eminently reasonable that loss of the leaving group should be the fast step—as in two step pathway (i) in [3.6] (p. 34)—for by losing $H^\oplus$ the intermediate regains the wholly aromatic condition of the benzene ring, with all that means in terms of stabilisation. This loss of $H^\oplus$ will be assisted by any anions present in solution, e.g. $HSO_4^\ominus$. Addition of such an anion to the cationic intermediate is also a possibility, but this would lead to a permanent forfeiture of aromatic character in the addition product with consequent loss of stabilisation:

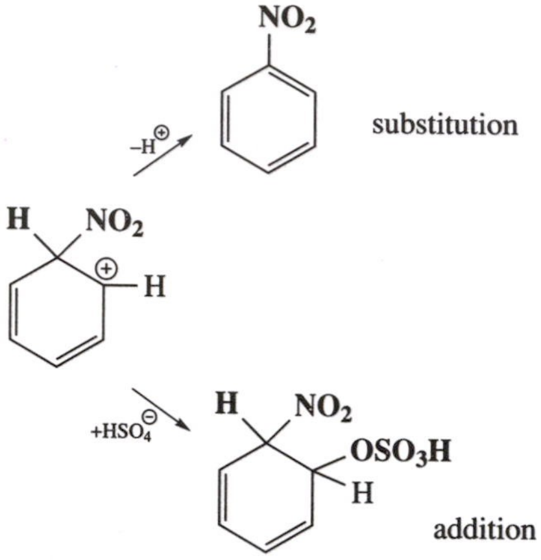

[3.8] SUBSTITUTION *VERSUS* ADDITION

Overall substitution thus leads to recovery of aromatic character, while overall addition—of any nucleophilic species available in the solution—leads to its permanent loss: a less desirable result in energy/stabilisation terms.

3.2.1.5 Evidence for formation of intermediates

Intermediates such as the one in [3.6] (p. 34) are generally referred to as **Wheland intermediates** or **arenium ions**, but the question then arises—as always with potential intermediates—as to whether there is any independent evidence for their existence. So far as nitration is concerned the answer would appear to be no, but in another electrophilic substitution reaction of aromatic species which we shall be considering below—the Friedel–Crafts reaction (**3.2.3**, p. 38)—it has proved possible to isolate a Wheland intermediate, and then demonstrate its subsequent conversion into the normal end product of the overall substitution reaction.

3.2.2 Halogenation

There are a number of other electrophilic substitution reactions of benzene, and of aromatic species in general, many of which are of considerable synthetic importance. These reactions commonly follow the two step (i) pathway [3.6] (p. 34) that we have suggested for nitration; the main question that remains to be decided about them is usually the exact nature of the attacking electrophile.

So far as electrophilic attack by halogens is concerned, benzene itself is found not to undergo substitution with chlorine, bromine or iodine by themselves (though Cl_2 and Br_2 can be made to <u>add</u> to benzene, under certain conditions (**6.3**, p. 99). Substitution can, however, be made to take place provided a suitable catalyst is present that is able to "step-up" the electrophilic character of the halogen molecule.

Classically the catalyst used was iron filings, though they do not actually function in that form: the halogen present, e.g. Br_2, first converts Fe into the corresponding Fe^{III} halide, e.g. $FeBr_3$, and it is this which acts as the catalyst:

$$Br-Br: \rightarrow FeBr_3 \longrightarrow \overset{\delta+}{Br}-Br \cdots \overset{\delta-}{FeBr_3}$$

[3.9] $FeBr_3$ "STEPS-UP" ELECTROPHILIC CHARACTER OF Br_2

The Fe atom in $FeBr_3$ is capable of accepting an electron pair into its outer electron shell (species able to accept electron pairs in this way are known as **Lewis acids**), and can thus form a complex with Br—Br in which one end of the bromine molecule has become $+^{vely}$ polarised, i.e. electron-deficient, and thus a more powerful electrophile:

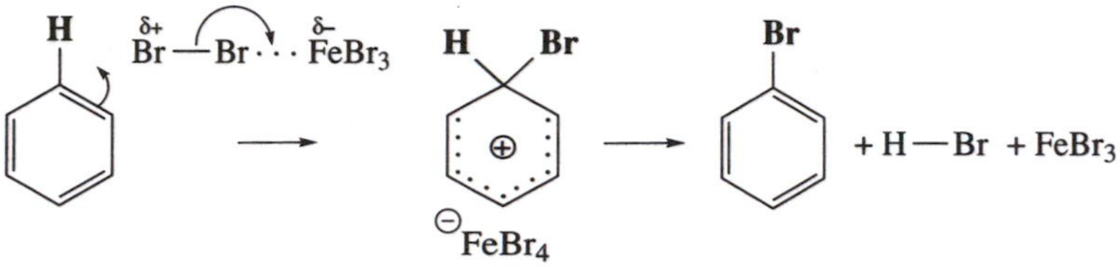

[3.10] BROMINATION OF BENZENE

The electrophilic end of the catalyst/Br_2 complex attacks the benzene ring to form an intermediate cation, and also the anion $FeBr_4^{\ominus}$; the latter is then able to assist in the removal of $H^{\oplus}$ from the cationic intermediate to form the end-product, bromobenzene. There are a number of different Lewis acids that can act as catalysts in a similar way, e.g. $AlCl_3$, and halogenation can also be effected by a number of halogen derivatives as well as by the halogens themselves.

Fluorination—with F_2 itself—is too vigorous to be of preparative value, as it results in break-down of the molecules being attacked. Iodine itself is not reactive enough to attack benzene, even with the assistance of a catalyst, but will attack more reactive aromatic species—such as phenol—without the need for any catalyst, as will the other halogens, e.g. bromine:

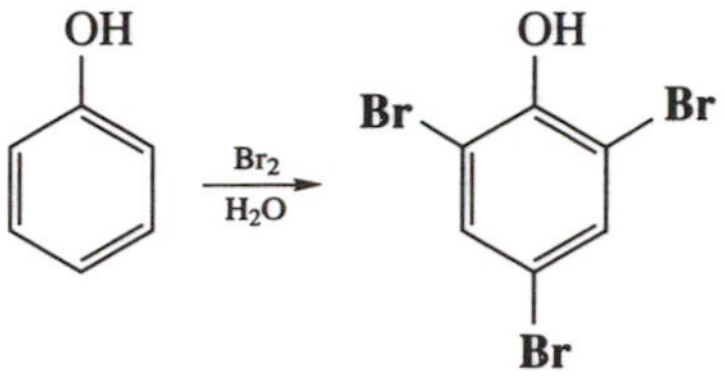

[3.11] BROMINATION OF PHENOL

We will consider the electrophilic substitution of other aromatic species— ones that are both more and less reactive than benzene itself—below (**3.2.5**, p. 41).

3.2.3   Friedel–Crafts reaction

This reaction involves the substitution of an H atom in aromatic species by an alkyl, R, or an acyl, RCO, group.

3.2.3.1   Alkylation

An alkyl halide, e.g. R—Cl, is itself polarised so that R constitutes an electrophilic "end" to the molecule:

$$R \rightarrow Cl \; \equiv \; R \; :Cl \; \equiv \; \overset{\delta+}{R} — \overset{\delta-}{Cl}$$

[3.12] ELECTROPHILIC "END" IN R—Cl

This polarisation is, however, not sufficiently pronounced to allow of attack on benzene by R—Cl alone, and—as with halogenation (**3.2.2**, p. 36)—a Lewis acid catalyst is also required. This results in a very similar pattern of reaction to that of halogenation, with a polarised complex as the attacking electrophile:

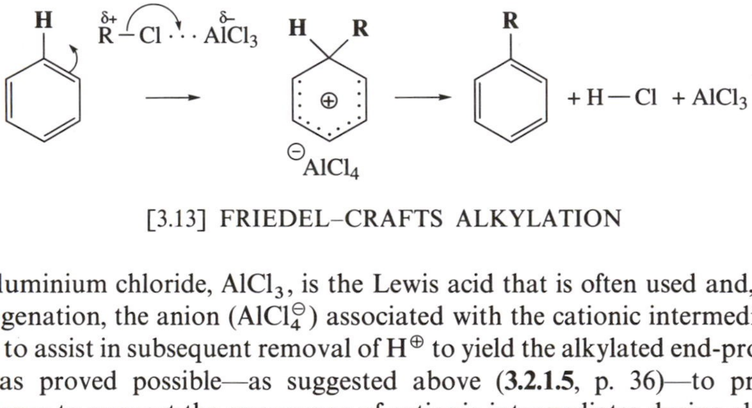

[3.13] FRIEDEL–CRAFTS ALKYLATION

Aluminium chloride, $AlCl_3$, is the Lewis acid that is often used and, as in halogenation, the anion ($AlCl_4^{\ominus}$) associated with the cationic intermediate is able to assist in subsequent removal of $H^{\oplus}$ to yield the alkylated end-product. It has proved possible—as suggested above (**3.2.1.5**, p. 36)—to provide evidence to support the occurrence of cationic intermediates during electrophilic substitution by actually isolating, and characterising, one in the course of an alkylation reaction (though not of benzene itself):

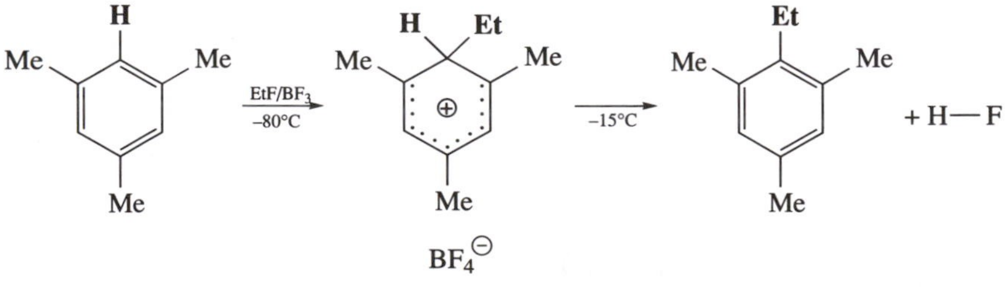

[3.14] ISOLATION OF AN INTERMEDIATE

The intermediate is an orange solid whose structure can be established spectroscopically; on being allowed to warm up above its melting point ($-15\,°C$), it is converted in essentially quantitative yield into the expected alkylated end-product. The stability of this particular intermediate, as part of an ion pair, is due in no small measure to the great stability of the anion, $BF_4^\ominus$.

Although in many alkylation reactions the attacking electrophile is the polarised complex, of R—Hal with a Lewis acid, that we saw above, if R is capable of forming a particularly stable carbocation, e.g. $Me_3C^\oplus$ (*cf.* [5.19], p. 76) from Me_3C—Br, then the attacking electrophile may well be the alkyl cation in an actual ion pair:

$$\overset{\delta+}{Me_3C}\!-\!\!\underset{\underset{\displaystyle\smile}{}}{\overset{\delta-}{Br}}\,,\ AlBr_3 \longrightarrow \overset{\oplus}{Me_3C}\ \ \overset{\ominus}{BrAlBr_3}$$

ion pair

[3.15] CARBOCATION AS ELECTROPHILE IN ION PAIR

One of the drawbacks to the use of Friedel–Crafts alkylation, as a preparative procedure, is that the product, C_6H_5R, is attacked by the electrophilic reagent more readily than is benzene itself; it is thus often difficult to stop at mono-alkylation, and unwanted poly-alkylated products are common (but see [3.18], p. 40). We shall be discussing below (**3.2.5.2**, p. 48) whether substituted derivatives of benzene, C_6H_5Y, react with electrophiles faster or slower than does benzene itself.

3.2.3.2 Acylation

Acylation follows the same general pattern as alkylation, but this time using an acyl halide, e.g. RCOCl, and a Lewis acid catalyst:

[3.16] FRIEDEL–CRAFTS ACYLATION

We find that in acylation the attacking electrophile can again be either a polarised complex or the acyl cation in an actual ion pair, depending on the

$$\overset{\delta+}{R\!-\!C}\!-\!\overset{\delta-}{Cl}\cdots AlCl_3 \qquad\qquad R\!-\!\overset{\oplus}{C}\ \ \overset{\ominus}{AlCl_4}$$
$$\ \ \ \ \|\qquad\qquad\qquad\qquad\qquad \|$$
$$\ \ \ \ O\qquad\qquad\qquad\qquad\qquad O$$

polarised complex   ion pair

[3.17] ATTACKING ELECTROPHILE IN ACYLATION

particular acyl halide and Lewis acid involved; but actual ion pairs are probably involved more often than in alkylation. It has proved possible to isolate the ion pair—$CH_3CO^{\oplus}BF_4^{\ominus}$—as a crystalline salt, and to show that this salt is itself capable of acetylating aromatic species very readily. The stability of this ion pair, which makes possible its isolation, stems again from the great stability of the $BF_4^{\ominus}$ anion in the ion pair (cf. [3.14], p. 38).

A larger proportion of Lewis acid catalyst is required in acylation than in alkylation, because the acylated product complexes (through the electrons on the O atom of its C=O group) with the Lewis acid, thereby preventing the latter from catalysing the acylation of further, as yet unreacted, starting material.

The end-product of Friedel–Crafts acylation, C_6H_5COR, undergoes less ready attack by electrophiles than does benzene itself, so there is no problem with poly-acylation comparable to the problem with alkylation. Indeed preparative alkylation is often carried out not directly, but *via* initial acylation followed by reduction (Clemmensen reaction) of the first-formed acyl product:

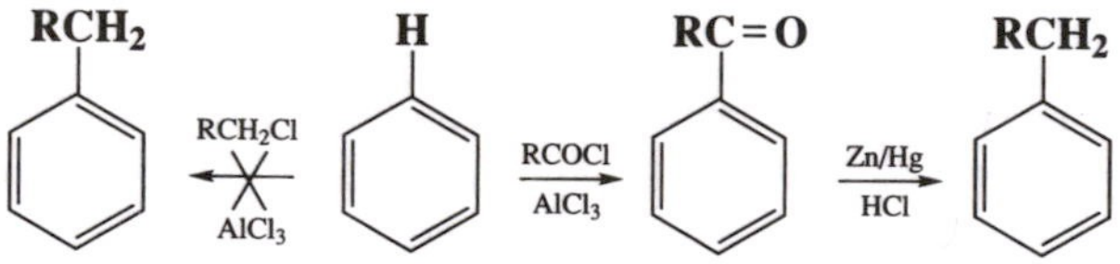

[3.18] ALKYLATION *VIA* ACYLATION/REDUCTION

3.2.4 Sulphonation

The sulphonation of benzene can be effected by using concentrated H_2SO_4 alone, but only at elevated temperature:

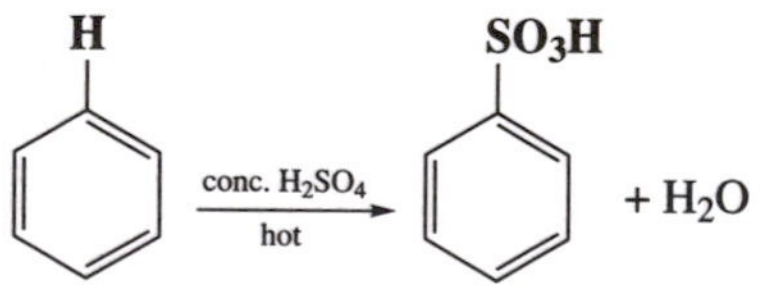

[3.19] SULPHONATION OF BENZENE

This reaction occurs very much more readily when the concentrated H_2SO_4 has some sulphur trioxide (SO_3) dissolved in it (this mixture is called oleum), and benzene can also be sulphonated by solutions of SO_3 in other (inert) solvents. It seems likely that in sulphonation SO_3 is always the effective electrophile, though it may sometimes be linked to a carrier molecule. The extent to which concentrated H_2SO_4 itself is capable of sulphonating aromatic species is due to the small concentration of SO_3 produced in the sulphuric acid through the equilibrium:

$$2H_2SO_4 \rightleftharpoons \underline{SO_3} + H_3O^{\oplus} + HSO_4^{\ominus}$$

[3.20] SO_3 FROM H_2SO_4

Sulphur trioxide is capable of acting as a powerful electrophile, through its sulphur atom, because this atom is highly $+^{vely}$ polarised:

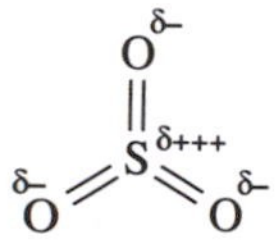

[3.21] SO_3 AS AN ELECTROPHILE

Sulphonation by SO_3 is believed to proceed according to the pathway:

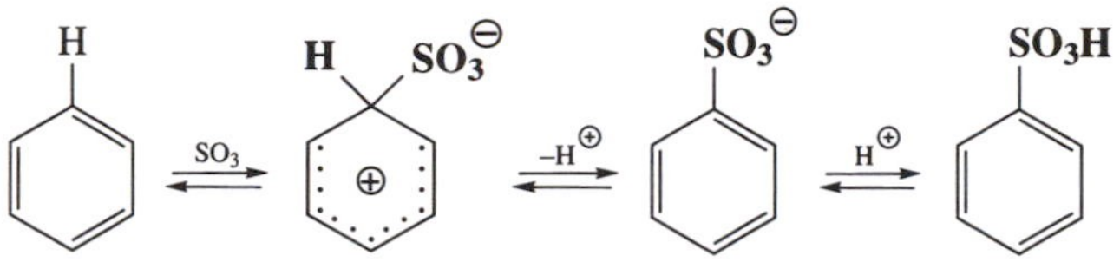

[3.22] SULPHONATION BY SO_3

As can be seen in [3.22], sulphonation is reversible—an unusual feature in an aromatic electrophilic substitution reaction—and an SO_3H group attached to an aromatic system may often be replaced by H on heating the sulphonic acid with steam.

3.2.5   Substitution in C_6H_5Y

In considering electrophilic attack on a benzene ring that already contains a substituent Y, i.e. C_6H_5Y, there are essentially two questions that need to be answered: (i) will substitution take place at positions *o*-(2-), *m*-(3-) or *p*-(4-) to Y, or at a mixture of all of them?

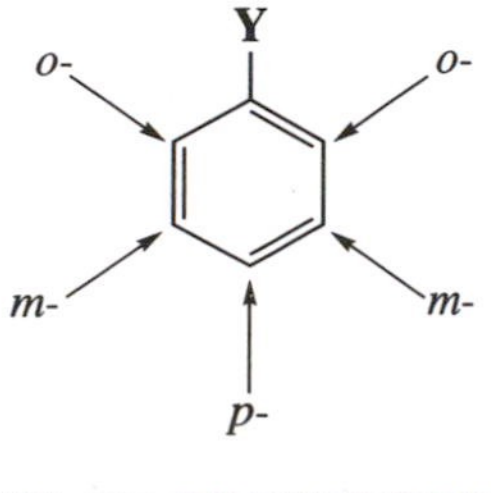

[3.23] POSSIBLE POSITIONS OF ELECTROPHILIC ATTACK ON C_6H_5Y

And (ii) will electrophilic attack on C_6H_5Y be faster or slower than on benzene itself under the same conditions?

3.2.5.1   Position of attack in C₆H₅Y

We might perhaps have expected to get attack on all three positions, obtaining the three possible products in statistically determined amounts, i.e. 40% *o*-, 40% *m*- and 20% *p*-. Although we might not have been unduly surprised to obtain rather less than 40% of *o*- product, as attack so close to the substituent Y could well have been subject to some steric interference.

This distribution of substitution product is **not**, however, what we actually obtain in the laboratory. In practice we obtain **either** a mixture of *o*- and *p*-substitution products, **or** the *m*-substitution product essentially alone: which pattern is followed is found to be governed by the character of the substituent **Y**. A particular Y substituent may thus be described as being either *o*-/*p*-**directing** or *m*-**directing**:

o-/*p* - directing	*m* - directing
CH₃	$\overset{\oplus}{N}R_3$
OH, OR	NO₂
NH₂, NR₂	CHO, RCO
Cl, Br, I	CO₂H, CO₂R
	SO₃H

[3.24] DIRECTING ABILITY OF THE SUBSTITUENT Y IN C₆H₅Y

3.2.5.1.1   o-/p-Direction

All *o*-/*p*-directing groups are found to have electrons available on the atom attached directly to the benzene ring, which they are able to share with the carbon atoms of the ring—they are **electron-donating** groups, e.g. CH₃, OCH₃:

[3.25] *o*-/*p*-DIRECTING GROUPS HAVE ELECTRONS TO SHARE WITH THE BENZENE RING

We have commonly written the cationic intermediates involved in electrophilic substitution of benzene in a form in which the +^{ve} charge is shown as being delocalised over the six-membered ring (**3.2.1.2**, p. 34). In fact the +^{ve} charge is not distributed <u>equally</u> over the five remaining unsaturated carbon atoms of the original benzene ring; this was demonstrated in spectroscopic studies on the intermediate involved in the reversible protonation of benzene (by HCl/AlCl₃):

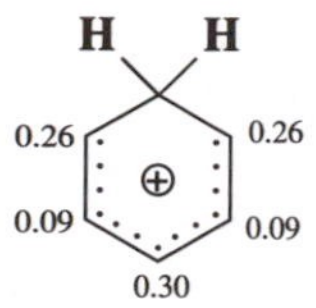

H   H

0.26   0.26
0.09 ⊕ 0.09
0.30

[3.26] UNEQUAL DISTRIBUTION OF $+^{VE}$ CHARGE IN CATIONIC
INTERMEDIATES

This unequal distribution of charge becomes more understandable when
we come to write out the several alternative "classical" (canonical) structures
(*cf.* [2.14] in **2.1.4**, p. 19) that each cationic intermediate could have in, for
example, the nitration of C_6H_5Y:

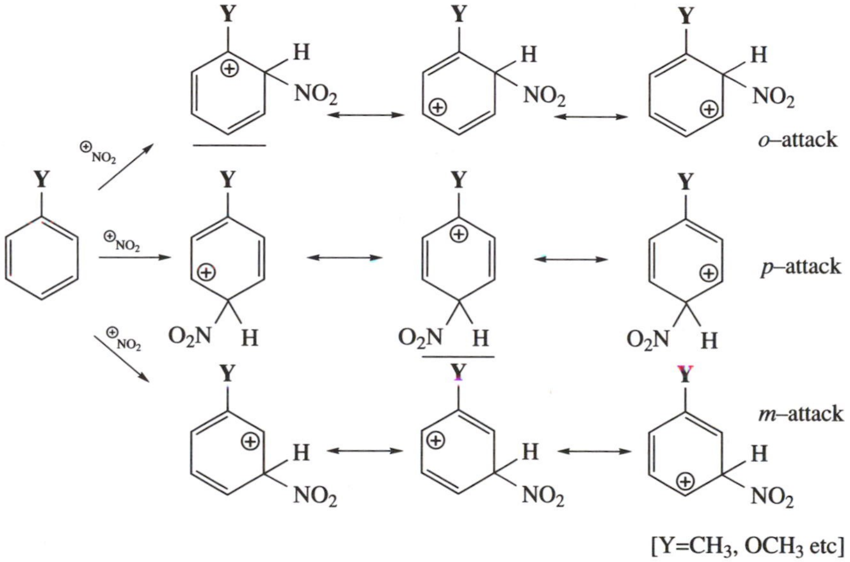

o–attack

p–attack

m–attack

[Y=CH$_3$, OCH$_3$ etc]

[3.27] ALTERNATIVE "CLASSICAL" STRUCTURES FOR
INTERMEDIATES

The actual electron distribution in each intermediate is really a composite
of three canonical structures (the double-headed arrows, ↔, represent the
essential equivalence of individual canonical structures), and cannot ade-
quately be portrayed in a single classical structure. Life is, however, too short
to write out several different structures every time we wish to represent an
intermediate, and the delocalised form is thus a convenient—if somewhat
inexact—compromise.

When we look closely in [3.27] at the canonical structures of the
intermediates involved in electrophilic attack at the *o*- and at the *p*-positions,
we see that in each case there is one structure (underlined) in which the $+^{ve}$

charge is located on the ring carbon atom that is bonded to the original, electron-donating, substituent Y:

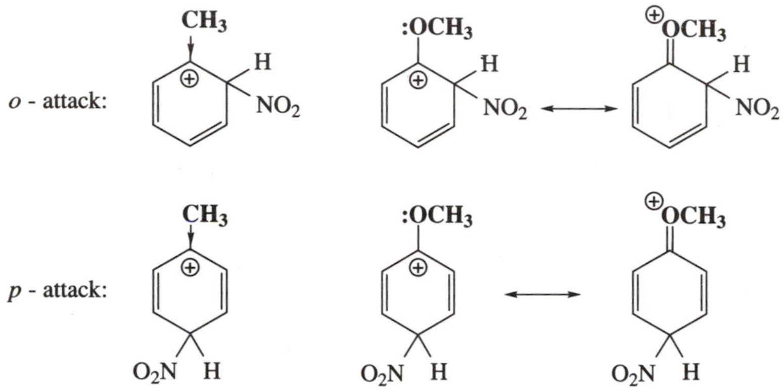

o - attack:

p - attack:

[3.28] FURTHER DELOCALISATION OF THE $+^{ve}$ CHARGE IN INTERMEDIATES FOR o- AND p-ATTACK

The $+^{ve}$ charge on these intermediates can thus be delocalised still further, with a consequent gain in stability. This, in turn, implies that these intermediates will be formed the more readily or, in other words, that reaction leading to o- and p-products will be easier and more rapid. As can be seen in [3.27] (p. 43), there is no corresponding canonical structure, with consequent stabilisation, for the intermediate involved in attack on the m-position, which will thus be slower than the more favoured, and hence preferential, attack on the o- and p-positions.

It is important to emphasise that o-/p-directing groups, such as CH_3 and OCH_3, do not operate—as the name perhaps implies—by prescriptively directing an attacking electrophile solely to the o- and p-positions: there is always competition (kinetic) between all three possible positions of attack, and the more rapid reactions win! Thus the presence of an o-/p-directing substituent does not specifically preclude the formation of any m-product, but the amount formed—if any—will only be small, because the reactions leading to the o- and p-products are so much faster.

3.2.5.1.2   m-Direction

By contrast, the groups that we listed in [3.24] (p. 42) as being m-directing are all **electron-withdrawing** groups. If we write out the canonical structures for the intermediates involved in the nitration of C_6H_5Y, where Y is a m-directing substituent such as $^\oplus NR_3$ or NO_2, we see that for o- and p-attack there is again, in each case, one structure (underlined) in which the $+^{ve}$ charge is located on the ring carbon atom that is bonded to the original, electron-withdrawing, substituent Y:

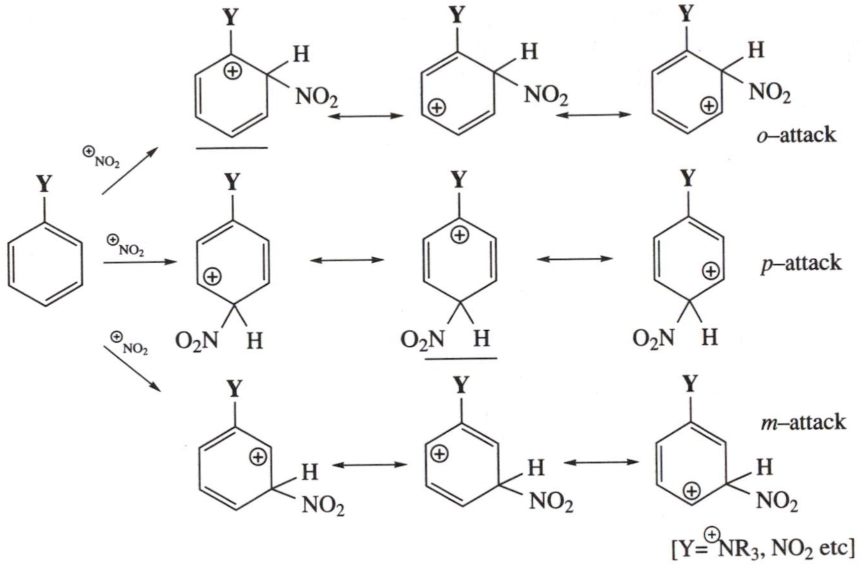

[3.29] ALTERNATIVE "CLASSICAL" STRUCTURES FOR
INTERMEDIATES

In both our examples of *m*-directing groups, the original substituent Y
itself carries a $+^{ve}$ charge, either real or formal:

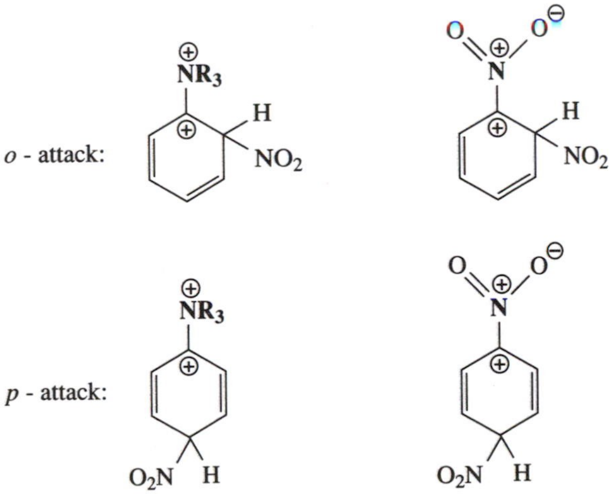

[3.30] DESTABILISATION OF INTERMEDIATES
FOR *o*- AND *p*-ATTACK

These canonical structures, in [3.30], thus carry $+^{ve}$ charges on adjacent
atoms: a highly unstable situation. The result is that the intermediates for

o- and *p*-attack are thus selectively **destabilised**, compared with the intermediate for *m*-attack in [3.29] (p. 45) which suffers no such destabilising juxtaposition. Attack on the *m*-position will thus—by default, as it were—be the fastest of the three, and is normally found to occur virtually exclusively because the destabilisation of the intermediates for *o*- and *p*-attack is so pronounced.

Exactly similar considerations apply even if Y doesn't carry an actual $+^{ve}$ charge; it will still be strongly electron-withdrawing, so that the carbon atom through which it is attached to the benzene ring, in the intermediates for *o*- and *p*-attack, will become $+^{vely}$ polarised, e.g. when Y is CHO, CO_2H, SO_3H, etc.

3.2.5.1.3   o-/p-Ratios

When considering those electrophilic substitution reactions of C_6H_5Y that led predominantly to the formation of a mixture of *o*- and *p*-products, we made no mention of what might influence the relative proportions in which these two products are obtained. We did, however, hazard a guess that attack on the position *o*- to the substituent Y might be somewhat more difficult than attack on the corresponding *p*-position, because of the possibility of steric hindrance by Y to the approach of the attacking electrophile, and also to the possibility of steric crowding in the intermediate for *o*-attack. This guess is borne out in practice by the observation that the relative proportion of *o*-product **decreases** as the **size** of both the substituent Y, and the attacking electrophile $E^\oplus$, **increases**:

	Y:	%o-	%p-		Reaction:	%o-	%p-
	CH_3	58	37		chlorination	39	55
increase in size of Y	$MeCH_2$	45	49	increase in size of $E^\oplus$	nitration	30	70
	Me_2CH	30	62		bromination	11	87
	Me_3C	16	73		sulphonation	1	99

[3.31] *o*-/*p*-RATIO: EFFECT OF SIZE OF Y AND OF $E^\oplus$

That size is not the sole consideration is, however, demonstrated by what is observed in the nitration of the four halogenobenzenes:

	Y:	%o-	%p-
	F	12	88
increase in size of Y	Cl	30	69
	Br	37	62
	I	38	60

[3.32] *o*-/*p*-RATIO: NITRATION OF HALOGENOBENZENES

Here the proportion of *o*-product actually **increases** as the **size** of Y increases. This reversal, compared with the effect of the four alkyl groups in [3.31] (p. 46), stems from the fact that the alkyl substituents differ very little in polarity from CH$_3$ to Me$_3$C, while the difference in polarity from F to I is marked indeed: F being very much more powerfully electron-withdrawing than I. Such electron withdrawal from the benzene ring will serve to inhibit attack by $^\oplus$NO$_2$, and this effect will be exerted to a greater extent on the *o*-positions, adjacent to Y, than on the much more distant *p*-position. This polar effect is sufficiently powerful to more than overcome the steric effect—I is much larger than F—that will still be operating in the opposite direction, cf. [3.31] (p. 46).

3.2.5.1.4   Ipso substitution

In our study of the reactions of electrophiles with C$_6$H$_5$Y, we have not yet considered whether attack could also occur at the carbon atom of the benzene ring to which the substituent Y is already attached:

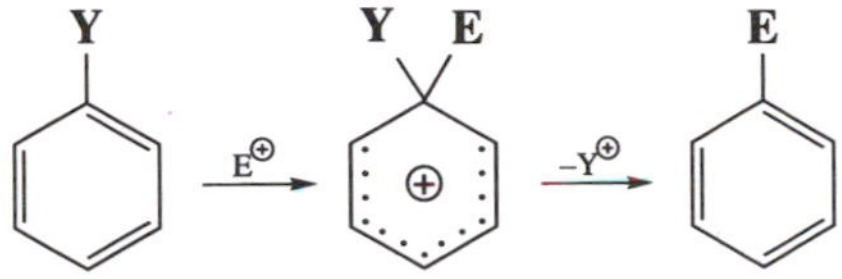

[3.33] *IPSO* SUBSTITUTION

This is known as *ipso* substitution, and requires Y$^\oplus$ to act as the leaving group in preference to H$^\oplus$: this is unlikely to happen all that often as H$^\oplus$ is an extremely good leaving group. We have, however, referred in passing to one example in **3.2.4** (p. 41), where it was mentioned that the sulphonation of benzene was reversible if the product sulphonic acid was treated with steam:

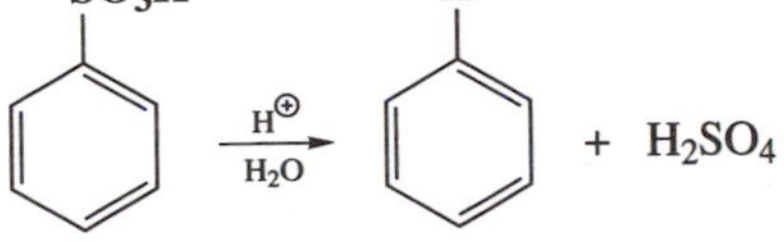

[3.34] REVERSAL OF SULPHONATION

Any feature in Y which serves to promote its stability as Y$^\oplus$ might be expected to promote its effectiveness as a leaving group; so one area in which we might look would be at alkyl substituents which can form reasonably stable carbocations—R$^\oplus$. Thus on nitration of the *p*-dialkylbenzene in [3.35] (p. 48), we obtain not only the expected *o*-nitration product, but also the *p*-nitro product in which the incoming NO$_2$ group has displaced one of the

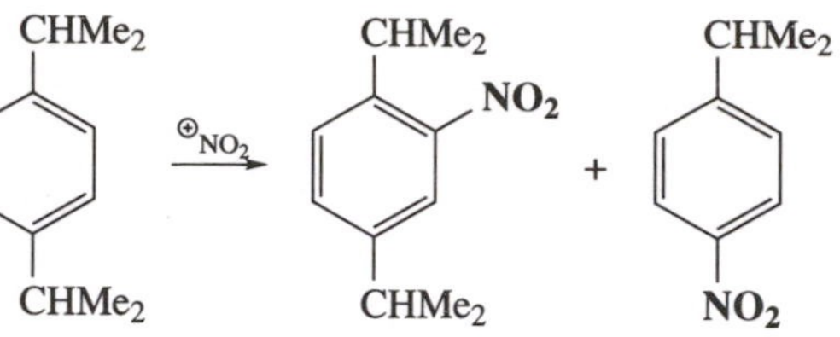

1   :   ≈ 5

[3.35] NITRATION OF A *p*-DIALKYLBENZENE

original Me_2CH substituents; we do indeed obtain five times as much of this unexpected product as of the orthodox *o*-nitro compound! $Me_2CH^{\oplus}$ is thus quite adequate as a leaving group in these circumstances. Hardly surprisingly, *ipso* attack is found to be promoted, as in [3.35], by any substituent—in this case Me_2CH—which serves to direct electrophilic attack towards the *ipso* position.

Occasionally it has proved possible to "trap" the cationic intermediate involved in *ipso* substitution by diverting it from the main reaction pathway, through converting it, in part at least, into an alternative end-product:

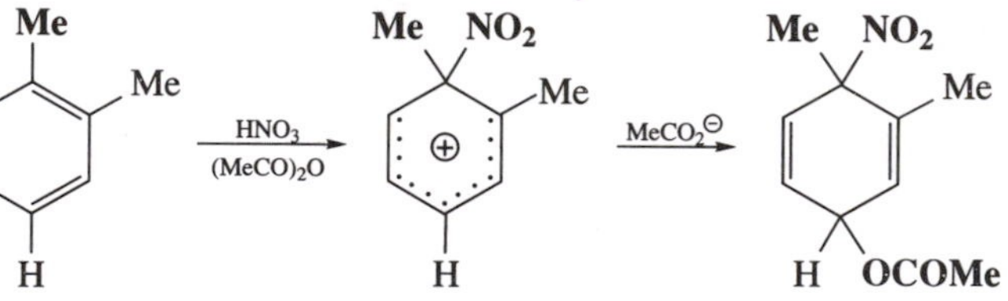

[3.36] "TRAPPING" OF THE INTERMEDIATE IN AN *IPSO* SUBSTITUTION

Thus in the nitration of 1,2-dimethylbenzene with HNO_3 in acetic anhydride $[(MeCO)_2O]$, formation of the effective electrophile ($NO_2^{\oplus}$) from HNO_3 also yields $MeCO_2^{\ominus}$, which is found—quite unexpectedly—to attack the cationic intermediate thereby forming a stable product from overall addition. The occurrence of the predicted intermediate for *ipso* attack is thus established through "freezing" its salient features in an alternative end-product.

Unlike the *o*-, *m*-, and *p*-substitution reactions of C_6H_5Y, which are of great preparative (synthetic) importance, *ipso* substitution is largely a potential preparative snag: something to watch out for when doing substitution reactions on aromatic compounds that already carry substituents.

3.2.5.2   Rate of attack on C_6H_5Y

It will now come as no surprise to hear that the nature of the substituent, Y, also controls the overall rate of attack on C_6H_5Y, as compared with that on C_6H_6. The general rule is that *o*-/*p*-directing (electron-donating) substituents cause attack to be faster than on benzene itself, while *m*-directing (electron-withdrawing) substituents cause attack to be slower, cf. [3.24] (p. 42): the former substituents are described as being **activating**, the latter as **deactivating**.

An explanation for these effects can readily be provided by comparing the cationic intermediates for electrophilic attack on C$_6$H$_5$Y, and on C$_6$H$_6$. Thus for an *o-/p*-directing (activating) substituent, e.g. Me or OMe, electron

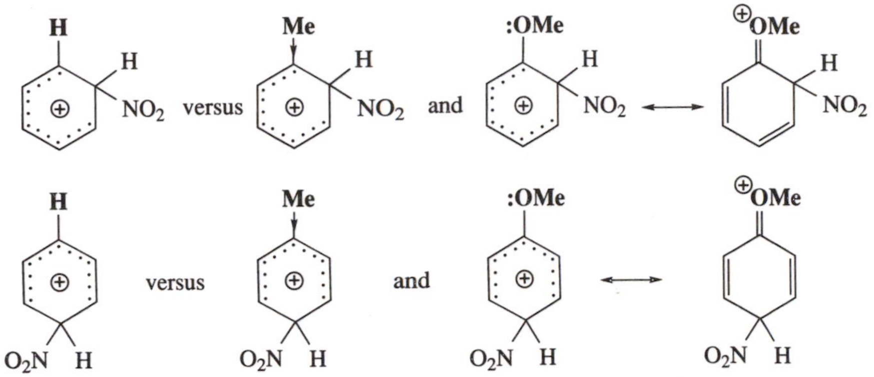

[3.37] COMPARISON OF INTERMEDIATES FOR NITRATION OF C$_6$H$_6$ AND C$_6$H$_6$Y, WHERE Y IS ACTIVATING

donation by Me, or by an electron pair on the oxygen atom of OMe, results in selective stabilisation of these intermediates (irrespective of whether attack is on the *o-* or the *p*-position) compared with the corresponding intermediate for attack on benzene itself. These intermediates are thus of lower energy than the one for similar attack on benzene, and are formed more rapidly.

By contrast, for a *m*-directing (deactivating) substituent, e.g. $^\oplus$NR$_3$, NO$_2$,

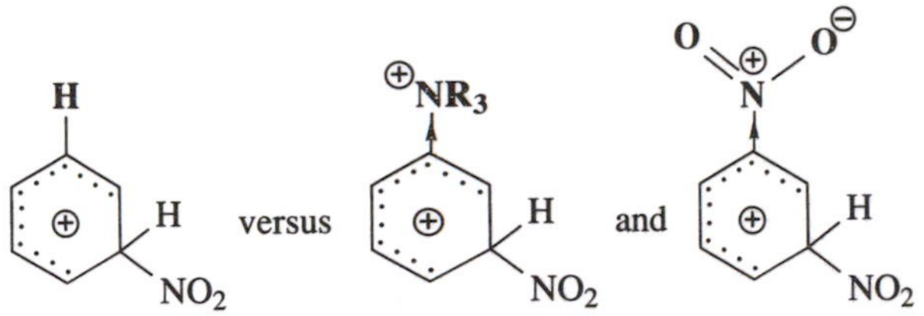

[3.38] COMPARISON OF INTERMEDIATES FOR NITRATION OF C$_6$H$_6$ AND C$_6$H$_5$Y, WHERE Y IS DEACTIVATING

electron-withdrawal by $^\oplus$NR$_3$, or by NO$_2$, results in selective <u>de</u>stabilisation of these intermediates compared with the corresponding intermediate for attack on benzene itself. These intermediates are thus of higher energy than the one for similar attack on benzene, and are formed more slowly.

Among the simple examples of Y there is, however, a significant anomaly: Cl, Br and I are *o-/p*-directing—as we have seen already—but they are, at the same time, <u>de</u>activating in that electrophilic attack on these halogeno-benzenes is found to be **slower** than similar attack on benzene itself.

We have already seen [3.28] (p. 44) how an electron pair on the oxygen atom of an OMe substituent is able to stabilise selectively the intermediates for *o*- and *p*-attack, and clearly the *o-/p*-directing halogens must be capable of doing the same:

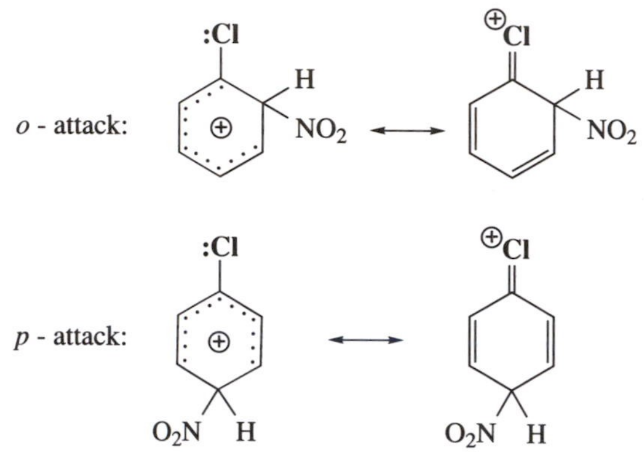

o - attack:

p - attack:

[3.39] *o-/p*-DIRECTION BY A HALOGEN SUBSTITUENT, e.g. Cl

There is, however, a difference in that Cl is found to be considerably more reluctant to share an electron pair with the attached benzene ring than is the O of OMe. This is demonstrated in the as yet unsubstituted molecules of chlorobenzene and methoxybenzene in [3.40], where we find that the **dipole moments** of the two molecules (reflecting the direction of electron-donation/withdrawal by the substituent) are in **opposite directions**:

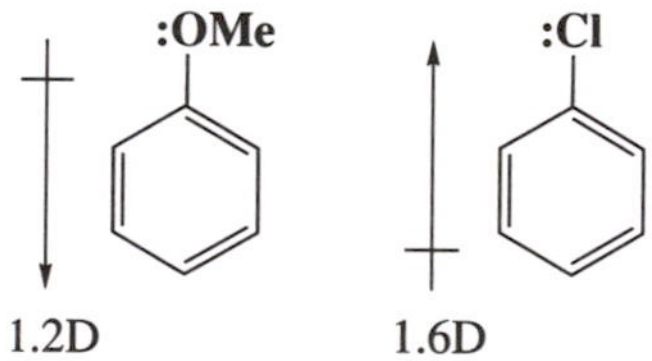

1.2D   1.6D

[3.40] DIPOLE MOMENTS OF C_6H_5OMe AND C_6H_5Cl

Thus while OMe, in C_6H_5OMe, is an overall electron-donor to the benzene ring, Cl, in C_6H_5Cl, is an overall electron-withdrawer.

When there is a $+^{ve}$ charge on the ring, as in the intermediates for *o*- and *p*-attack, the pull on the electron pairs on the chlorine and oxygen atoms is greatly increased, resulting in selective stabilisation of the relevant intermediates in each case. The response by Cl is considerably less than that by oxygen, however, with the result that there is still not enough net electron-donation to the ring to make attack on chlorobenzene faster than on benzene itself.

3.3 SUMMARY

Substitution at a **saturated** carbon atom by **electrophiles**—electron-deficient reagents—is not a reaction of great significance, but electrophilic substitution at an **unsaturated** carbon atom is; particularly when the unsaturated carbon atom is part of a benzene ring—**aromatic** substitution.

The example of aromatic substitution that has been most studied is **nitration**. Consideration is given to the nature of the attacking electrophile in nitration, to the rate equation for the reaction, to the different reaction pathways that are compatible with such an equation, and to how we can decide between them. Finally, an explanation is given of why electrophilic attack on aromatic systems leads to overall <u>substitution</u>, rather than to the overall <u>addition</u> that might perhaps have been expected.

Attack on aromatic systems by other electrophiles—halogenation, Friedel–Crafts alkylation and acylation, and sulphonation—are also described. These are, in general, found to follow reaction pathways essentially analogous to that for nitration, the point at issue commonly being the actual nature of the electrophile involved in the reaction.

Consideration is then given to electrophilic substitution on a benzene ring that already contains a substituent, i.e. C_6H_5Y; and to how the substituent, Y, influences both the <u>position</u> of attack on C_6H_5Y (*o*-, *m*-, or *p*-), and the <u>rate</u> of attack on it, compared with the rate of similar attack on C_6H_6. An explanation of both influences is provided in terms of the selective stabilisation of the relevant cationic intermediates involved. Reference is also made to electrophilic attack on C_6H_5Y in which the original substituent, Y, rather than H is replaced—*ipso* substitution.

4

Radical substitution

We have already seen (**1.6.3**, p. 6) that the very low polarity of C—H bonds results in their being largely insusceptible to attack by either nucleophiles or electrophiles, even under fairly vigorous conditions; C—H bonds are, however, found to react readily with species that have a single (unpaired) electron in their outer shell, namely **radicals**.

4.1 FORMATION OF RADICALS

Irradiation of a molecule of chlorine, Cl_2, with light of suitable wavelength (**1.6.3**, p. 6) causes breaking (photolysis) of the Cl—Cl bond, resulting in the formation of two chlorine atoms or radicals, Cl· (the movement of a single electron is commonly represented by an arrow having only a single barb, in contrast to the doubly barbed arrows used to represent the movement of electron pairs):

$$\overset{\frown}{Cl} : Cl \xrightarrow{\text{light}} Cl \cdot \quad \cdot Cl$$

[4.1] PHOTOLYSIS OF A Cl—Cl BOND

The high reactivity of Cl· stems from its unpaired electron seeking another electron, with which it can achieve the more stable state of an electron pair.

Radicals can also be generated in a number of other ways—through heating, for example. Thus the general breakdown of organic compounds at high temperatures occurs largely *via* the agency of radical formation and reaction, under these extreme conditions. The **thermolysis** (fission by heat) of metal alkyls can, however, be effected at much lower temperatures:

$$Pb(CH_2Me)_4 \xrightarrow[\text{heat}]{} Pb + 4\ MeCH_2\cdot$$

[4.2] THERMOLYSIS OF LEAD TETRAETHYL

Thus such lead alkyls were put into petrol to act as <u>anti-knock</u> agents. The Pb—C bond is a weak one, and breaks quite easily; the **alkyl** radicals, e.g. $MeCH_2\cdot$, so produced combine with some of the radicals being formed from the heat breakdown of the petrol hydrocarbons. This ensures that the overall oxidative breakdown of these hydrocarbons occurs in a smooth and controlled way, thus avoiding the occurrence of the too vigorous sites of over-rapid reaction that constitutes knocking (cf. **4.2.1**, p. 54). In these days of lead-free petrol, knocking is now avoided in other ways. Combustion in general, and the combustion of petroleum hydrocarbons in particular, is—on a tonnage basis—by far the most important radical reaction of all!

Another thermolytic reaction that leads to the generation of radicals, at quite low temperatures, is the fission of peroxides to form **alkoxy** radicals, $RO\cdot$:

$$RO : OR \xrightarrow[\text{heat}]{} RO\cdot\ \cdot OR$$

[4.3] THERMOLYSIS OF PEROXIDES

There are also a number of ways in which radicals can be generated through oxidation/reduction reactions that involve the transfer of a single electron. An example is the use of $Fe^{2\oplus}$ to generate hydroxyl radicals, $\cdot OH$, from hydrogen peroxide, H_2O_2:

$$HO : OH + Fe^{2\oplus} \longrightarrow HO\cdot + {}^{\ominus}OH + Fe^{3\oplus}$$

[4.4] GENERATION OF HYDROXYL RADICALS FROM H_2O_2

Once formed, the major reaction of such radicals is the **abstraction** of H from C—H bonds.

4.2 SUBSTITUTION REACTIONS

You may remember (**1.6.3**, p. 6) that methane, CH_4, which contains <u>only</u> C—H bonds, does not react with chlorine in the dark, but if light (of suitable wavelength) is shone on the mixture then reaction occurs so rapidly that explosion may well occur.

4.2.1   Halogenation

Light **photolyses** the chlorine molecule ([4.1]. p. 52) to form two chlorine radicals, Cl·, each of which is then able to react with a molecule of methane:

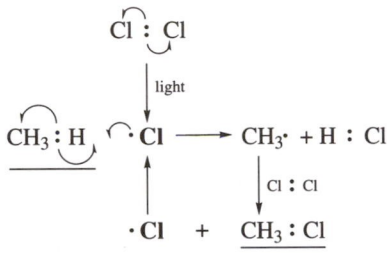

[4.5] REACTION OF CH_4 WITH Cl·

The highly reactive Cl· abstracts a hydrogen atom, H·, from the methane molecule to form H:Cl, thereby producing the very reactive methyl radical, CH_3·. In turn, this methyl radical abstracts a chlorine atom from a further molecule of chlorine to form the expected end-product, CH_3:Cl, also producing, at the same time, a new chlorine radical. This new chlorine radical is now able to repeat the whole process of turning a further molecule of CH_4 into CH_3—Cl, while again yielding a new chlorine radical, which can repeat the whole cycle yet again, and so it goes on.

A repeating cycle of reaction like this is called a **chain reaction**. In this particular case, each "packet" of light (photon) absorbed produces two chlorine radicals, **each** of which is found to set off a sequence leading to the formation of approximately two million molecules of substitution product, CH_3—Cl! As the light shone on the reaction is made up of an enormous number of individual photons, it is no longer a surprise that the rate of the reaction should be rapid enough to lead to explosion. Although the overall reaction is one of <u>substitution</u> it has in fact been achieved *via* two successive <u>abstractions</u> of atoms.

An obvious question is why such a chain reaction, once it has been initiated by a chlorine radical, should not continue until there are no further molecules of methane left, instead of stopping after a mere million repetitions! The answer is that reactive radicals can, of course, react with each other, e.g. Cl· + ·Cl → Cl_2, CH_3· + ·CH_3 → CH_3—CH_3, CH_3· + ·Cl → CH_3—Cl, thereby terminating <u>two</u> reaction chains in each case, as no new radical is generated to carry on the process. The number of radicals present in the system at any one time is, however, so vanishingly small that collision between <u>two</u> of them is likely to be an extremely rare event. A particular reaction cycle is thus able to repeat many, many times, before it is eventually terminated by the fortuitous collision of two of these very scarce radicals.

4.2.1.1   Attack at primary, secondary and tertiary hydrogen

We have, as yet, considered only the chlorination of methane, CH_4, in which all four C—H bonds are identical; but in many simple hydrocarbons, e.g. $CH_3CH_2CH_3$, all the C—H bonds are not identical, and are in different local environments to the C—H bonds in CH_4. Thus if we compare the C—H bonds in hydrocarbons which have primary, secondary and tertiary H atoms, we find a small but significant difference in reactivity between

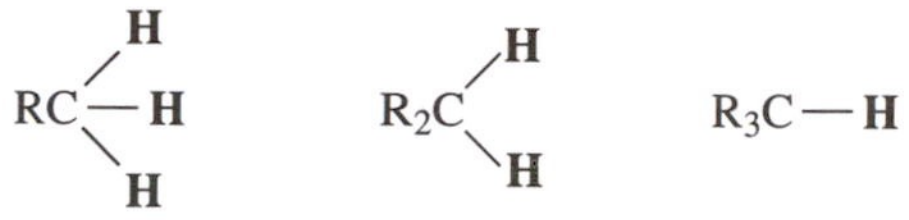

RC—H (primary H, 1)   R₂C (secondary H, 4.4) R_3C—H (tertiary H, 6.7)

primary H — 1
secondary H — 4.4
tertiary H — 6.7

[4.6] RELATIVE REACTIVITY OF PRIMARY, SECONDARY AND TERTIARY H TOWARDS Cl·

them: tertiary H being the most reactive of the three. This difference in reactivity reflects the relative strength of the three different C—H bonds, and also the relative stability of the three different alkyl radicals that will be formed by H· abstraction. These two properties are found to follow the orders shown below:

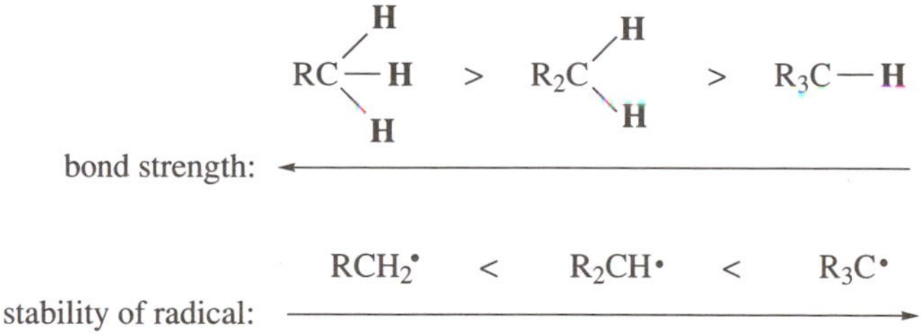

RC—H > R₂C > R_3C—H

bond strength: ⟵————————————————————

$RCH_2^{\bullet}$ < $R_2CH\cdot$ < $R_3C\cdot$

stability of radical: ————————————————————⟶

[4.7] BOND STRENGTH/RADICAL STABILITY FOR PRIMARY, SECONDARY AND TERTIARY SITES

The strength of the C—H bond **decreases** as we go from RCH_3 to R_3CH, while the stability of the radical formed (through H· abstraction) **increases** across the three hydrocarbons. These two effects thus reinforce each other to make H-abstraction progressively easier across the series: H· abstraction is the step that normally dictates the rate of overall chlorination.

If we hope to forecast the composition of the product that would be obtained from the chlorination of a hydrocarbon such as $(CH_3)_3CH$, we have to take into account not only the difference in reactivity of the two different types of C—H bond that are involved, but also a **statistical** effect: there are **nine** primary C—H bonds to only **one** tertiary C—H. In theory, we could thus calculate the ratio in which the two mono-chloro products should be obtained; but when the reaction is actually carried out in the laboratory the result only roughly parallels what we had calculated:

$$CH_2 — Cl$$
$$(CH_3)_2CH \qquad (CH_3)_3C — Cl$$

observed:	65%	35%
calculated:	57%	43%

[4.8] PROPORTIONS OF MONO-CHLORO PRODUCTS FROM
CHLORINATION OF $(CH_3)_3CH$

As so often happens, things are not quite as straightforward as we might perhaps have imagined!

4.2.1.2   Effect of halogenating agent

There is found to be a considerable difference between the halogens in their relative abilities to effect radical attack on the C—H bond. Thus attack by fluorine, F_2, takes place without needing specific generation of F· by photolysis or other means; this reflects the fact that the F—F bond is very weak, and breaks to form two F· atoms extremely easily. The reaction of fluorine itself with hydrocarbons is found to be so vigorous that it often results in a general breakdown of the molecule being fluorinated. Direct fluorination is commonly of little preparative value, but indirect methods have been devised.

Direct chlorination we have already discussed in some detail. In bromination, the bromine radical, Br·, is found to be considerably less reactive than Cl·, which results in much greater susceptibility to differences in reactivity between differently situated C—H bonds:

primary H	secondary H	tertiary H
$RC{<}^{H}_{H}{—}H$	$R_2C{<}^{H}_{H}$	$R_3C — H$
1	80	1600

[4.9] RELATIVE REACTIVITY OF PRIMARY, SECONDARY AND
TERTIARY HYDROGENS TOWARDS Br·

The greater **selectivity** of Br·, compared with Cl·, is very marked, and this is reflected in the fact that bromination of $(CH_3)_3CH$ results in the formation of $(CH_3)_3C$—Br only (*cf.* Cl· in [4.6], p. 55). Bromination is thus often more useful preparatively than chlorination, as its greater selectivity leads to "cleaner" products. The iodine atom (radical), I·, is so poor an abstractor of H· that direct iodination of hydrocarbons with I_2 is not normally possible.

Chlorination with Cl$_2$ tends to give mixtures of mono-chloro products, as we have seen, but its lack of selectivity often leads, in addition, to further chlorination of the initial mono-chloro products, thereby producing even more complex product mixtures. These problems can be largely overcome by the use of agents more selective than chlorine itself; chlorination can then become a useful preparative procedure.

A good example is the use of alkyl hypochlorites, e.g. Me$_3$COCl, which require the introduction of a suitable radical, Ra·, to start the reaction off (**initiation**):

initiation: Ra·   Cl : OCMe$_3$ ⟶ Ra : Cl + ·OCMe$_3$

   R : H   ; OCMe$_3$ ⟶ R· + H : OCMe$_3$

          Cl : OCMe$_3$

   ·OCMe$_3$ +   R : Cl

[4.10] CHLORINATION WITH ALKYL HYPOCHLORITES

The introduced initiator radical, Ra·, abstracts the chlorine atom from a molecule of alkyl hypochlorite to form the alkoxy radical, Me$_3$CO·. This can then abstract a hydrogen atom from the hydrocarbon, R—H, to form the radical, R·. R· can, in turn, abstract the chlorine atom from a further molecule of alkyl hypochlorite to yield the looked-for chloro product, R—Cl, plus a further alkoxy radical, Me$_3$CO·, which can initiate a further cycle of chlorination: we have indeed set up a chain reaction, very similar to the one we first saw in [4.5] (p. 54).

The reason that chlorination with an alkyl hypochlorite is preparatively useful, in cases where direct chlorination with Cl$_2$ is not, stems from the fact that the species that effects the vital step of H· abstraction (which controls the overall reaction) is not Cl· but Me$_3$CO·. This radical is very much more selective in which H atom it will abstract from the hydrocarbon undergoing chlorination, thereby avoiding the formation of a mixture of different chloro- products.

4.2.1.3   Effect of an adjacent double bond

When an alkyl group, e.g. CH$_3$, is adjacent to a double bond, as in CH$_2$=CH—CH$_3$, we might well expect attack by halogens to result in addition to the double bond (*cf.* **1.2**, p. 2) rather than abstraction of H· from the alkyl group. Addition of halogen radicals is indeed observed (**6.1.1**, p. 87), but is found to be reversible: at higher temperatures, or with low concentrations of halogenating agent, H· abstraction is found to take place preferentially, leading overall to halogenation of the CH_3 group:

allyl radical: CH_2=CH—CH_2· ⟷ ·CH_2—CH=CH_2

·CH_2—CH=CH_2

CH_2=CH—CH_2 : H ⌒ ·Cl ⟶ CH_2=CH—CH_2· + H : Cl

Cl : Cl

·Cl + CH_2=CH—CH_2 : Cl

[4.11] STABILISATION OF A RADICAL BY AN ADJACENT DOUBLE
BOND IN THE CHLORINATION OF CH_2=CH—CH_3

Similar halogenation can also take place when the CH_3 group is attached to a benzene ring (cf. [6.18], p. 99).

The preferred abstraction of H· is due in part to the C—H bonds in such a methyl group being weakened slightly, but more particularly to stabilisation of the developing (allyl) radical through delocalisation of its unpaired electron by the adjacent double bond. Thus chlorination of CH_2=CH—CH_3 with Cl_2 at 450 °C (Cl· is produced from Cl_2 by thermolysis at this temperature) leads to the formation of CH_2=CH—CH_2Cl only. The advantage of high temperature is that at 450 °C the addition of chlorine to the double bond is reversible, while the chlorination of CH_3 is not. Alkyl hypochlorites, e.g. Me_3COCl as in [4.10] (p. 57), because of their high selectivity are particularly good at effecting chlorination of such unsaturated hydrocarbons.

There is a particularly effective reagent used preparatively for the bromination of such positions, namely N-bromosuccinimide:

[4.12] N-BROMOSUCCINIMIDE

Initially there was some disagreement over how this reagent actually worked, but there now seems to be little doubt that it acts by providing a regular supply of bromine in extremely low concentration. There is generally a very small amount of Br_2 or HBr present in N-bromosuccinimide (arising from its decomposition in the air); either will react with a suitable initiator radical—introduced into the system to set the reaction off—to generate Br·:

initiation: Br$_2$ or HBr + Ra· $\longrightarrow$ Br·

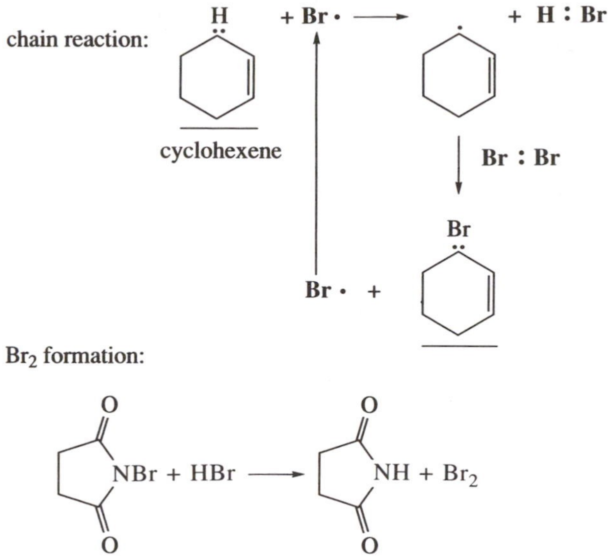

chain reaction:

Br$_2$ formation:

[4.13] BROMINATION BY N-BROMOSUCCINIMIDE

This bromine radical abstracts H· from the allylic position (the carbon atom adjacent to the double bond) of an unsaturated hydrocarbon, e.g. cyclohexene, to yield a cyclohexenyl radical plus HBr. This molecule of HBr reacts, in turn, with N-bromosuccinimide to form Br$_2$ which can then react with the cyclohexenyl radical to form the product bromo-compound plus a bromine radical, which can initiate a further cycle of bromination. A chain reaction is thus set up, but is only sustained by regular recourse to N-bromosuccinimide, which thereby keeps the concentration of Br· extremely low and so controls the overall reaction.

4.2.2   Autoxidation

As well as undergoing oxidative destruction at high temperature, in combustion, organic compounds may also undergo slower oxidative attack at more moderate temperatures; this is known as **autoxidation** and, like combustion, is a radical reaction. Both depend on the fact that the oxygen molecule contains <u>two</u> unpaired electrons—that it is a <u>di</u>radical, albeit a fairly unreactive one. Because of this unreactivity, the oxygen molecule itself is not usually capable of initiating autoxidation, which normally requires the introduction (*via* photolysis or other means) of a suitable initiator radical, Ra·, to set the process off:

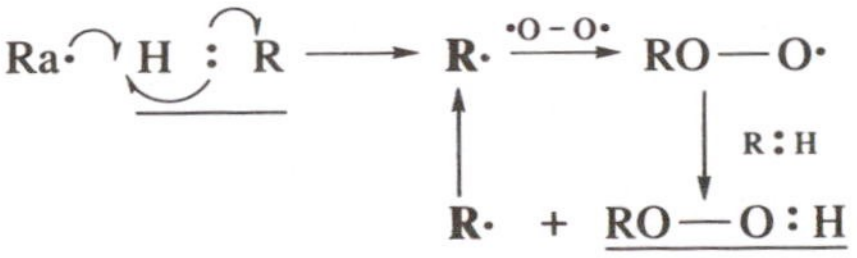

[4.14] AUTOXIDATION CHAIN REACTION

The initiator radical, Ra·, abstracts an H· atom from R—H, and the resultant alkyl radical, R·, combines with O_2 to form a peroxy radical, ROO·; this is slightly more reactive than ·O—O·, and is thus able to abstract an H· atom from R—H to form the end-product—a **hydroperoxide**, ROOH. A new alkyl radical, R·, is formed at the same time which can then repeat the autoxidation cycle, thus establishing the familiar chain reaction.

Peroxy radicals are usually not very reactive, and are thus highly selective in the type of C—H bond from which they are capable of abstracting an H· atom. Thus they tend to attack tertiary, allylic and benzylic (adjacent to a benzene ring) C—H bonds most readily:

H	H	H_2C—H
tertiary	allylic	benzylic

$CH_2 = CH — CH_2$

[4.15] SELECTIVITY OF ATTACK BY ROO·

These positions are attacked preferentially (as we have already seen, e.g. [4.7], p. 55, and [4.11], p. 58) because their C—H bonds are rather weaker, and the radicals produced from them—by H· abstraction—are stabilised through delocalisation.

In practical terms there are both advantages and disadvantages in autoxidation, which is going on all around us whether we like it or not. Thus it is involved in the hardening of paints and varnishes, where autoxidation of some of the unsaturated compounds that these contain leads to the formation of a protective surface film. This results from breaking of the weak O:O bond in the first formed hydroperoxides, RO:OH, thus forming new radicals, RO· and ·OH. These are reactive enough to initiate polymerisation of the unsaturated molecules in the paint or varnish, thus converting an initially liquid layer into a hard film. We shall be considering radical induced polymerisation in detail below (**6.2.3**, p. 96).

The deleterious effect of autoxidation is seen in many of the "ageing" processes that occur in organic compounds which, on standing in air and light, undergo photo-initiated oxidation: fats going rancid, rubber perishing, some plastic materials deteriorating; peroxy radicals are also believed to be capable of inducing damage in DNA. Though these ageing processes are commonly initiated by radicals produced through photolysis, they can also be initiated by the presence of trace metals in the form of ions (cf. [4.4], p. 53), and by other means.

It is possible, in part at least, to protect organic compounds from autoxidation by adding to them small amounts of compounds which are themselves known to react particularly readily with radicals. Such substances commonly possess H atoms that are readily abstracted by any radicals that are already present in the system; the new radicals produced by this H·

abstraction are, however, such as not to be reactive enough to set off a chain reaction. A number of different phenols, and aromatic amines, are used as radical "traps" in this way: in this particular context of inhibiting autoxidation, they are known as **anti-oxidants**.

4.2.3 Aromatic substitution

We might well expect benzene and other aromatic systems, because of their apparent unsaturation, to undergo addition when attacked by radicals, and this can indeed be made to occur under appropriate conditions (**6.3**, p. 99). Just as unsaturated species such as $CH_2{=}CH{-}CH_3$ could be made to undergo overall substitution ([4.11], p. 58), it is also possible to effect the substitution of benzene with radicals, as well as with nucleophiles (**2.2.1**, p. 25) and electrophiles (**3.2**, p. 32). Such overall radical substitution is believed to take place by a two step pathway, similar to those we have already seen operating in the attack of other reagents on benzene:

intermediate

[4.16] PATHWAY FOR RADICAL SUBSTITUTION OF BENZENE

The delocalised radical intermediate involved does not lose H· spontaneously, but requires H· abstraction by any of the radicals present in solution. Such abstraction is not difficult, however, and it is the initial attack by the "substituting" radical on the aromatic system that is found to be the slow step which controls the overall rate of reaction. It is worth emphasising that, unlike the radical substitution reactions that we have seen up to now, this is not a chain reaction.

A significant difference from electrophilic substitution is that radical attack on C_6H_5Y is always found to be faster than on C_6H_6 itself, irrespective of whether Y is activating (electron-donating) or deactivating (electron-withdrawing):

Y:	H	MeO	Cl	Br	CH₃	CN	NO₂
relative rate:	1.0	1.2	1.4	1.7	1.7	3.7	4.0

[4.17] RELATIVE RATES OF ATTACK ON C_6H_5Y BY $C_6H_5\cdot$ (Ph·)

Any substituent, irrespective of its nature, must thus be able to stabilise the intermediate in this reaction, thereby making attack on C_6H_5Y easier than on C_6H_6 itself.

It therefore comes as no surprise to find that radical aromatic substitution is not susceptible to any *o-/p-* or *m*-directing effects of Y either, unlike aromatic substitution by electrophiles. Attack usually takes place on all three positions, with *o*-attack commonly predominating. Thus for attack by $C_6H_5 \cdot$ (**4.2.3.1** below) on C_6H_5Y we find:

Y:	%*o*-	%*p*-	%*m*-
NO_2	63	10	27
Me	61	16	23
Cl	59	25	16
MeO	52	30	18

[4.18] POSITION OF RADICAL ATTACK ON C_6H_5Y BY $C_6H_5 \cdot$ (Ph·)

Two of the many different aromatic substitution reactions by radicals warrant some further study.

4.2.3.1   Phenylation

This, of all the reactions in this category, has received by far the most detailed study. The $C_6H_5 \cdot$ (Ph·) radicals required in this reaction may be generated in a number of different ways, but one of the most convenient is the decomposition of benzoyl peroxide (PhCOO—OOCPh) at relatively low temperature (80°C):

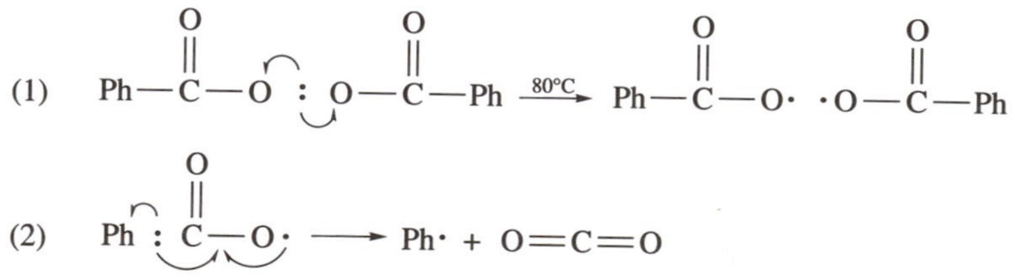

[4.19] GENERATION OF Ph· FROM (PhCOO)$_2$

The benzoyloxy radical, $PhCO_2 \cdot$ is formed very easily because the O—O bond in benzoyl peroxide is an extremely weak one, and may well not require even as high a temperature as 80 °C for fission to occur. This temperature does, however, ensure that the benzoyloxy radical is decarboxylated as soon as it is formed, thus avoiding it too being involved as a potential substituting species in the main reaction.

The expected end-product from the reaction of Ph· with benzene is, of course, biphenyl:

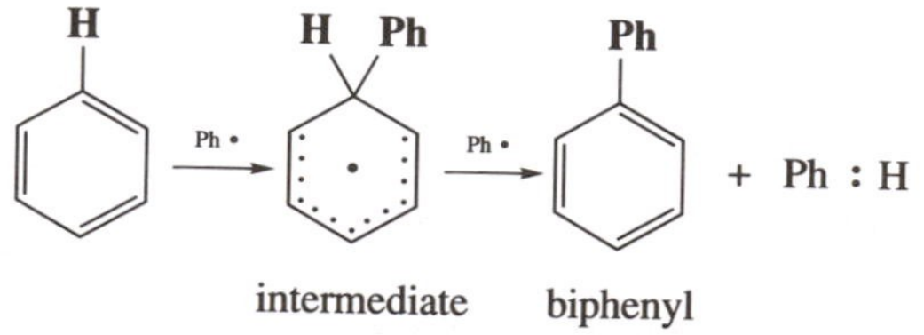

intermediate   biphenyl

[4.20] FORMATION OF BIPHENYL FROM Ph·/C_6H_6 REACTION

This is not, however, the only product that is obtained. The radical intermediate—because it is not caught up in an inexorable chain reaction—is found to be capable of reacting with itself:

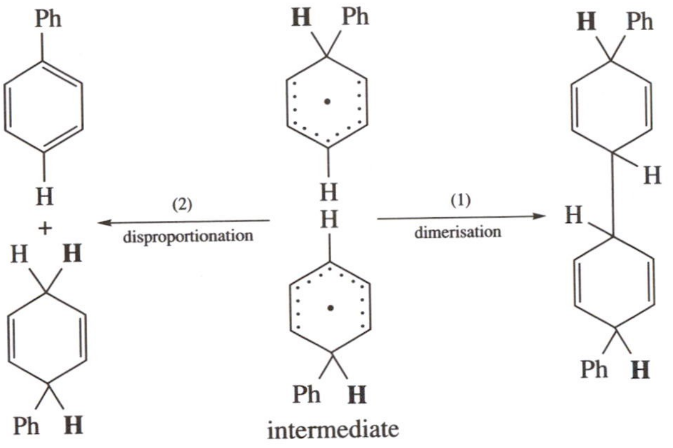

[4.21] TWO MOLECULES OF RADICAL INTERMEDIATE REACTING WITH EACH OTHER

In reaction (1) two molecules of radical intermediate have paired their single electrons, through the *p*-positions of their benzene rings, to yield a "doubled-up" product: this is known as **dimerisation**. In reaction (2) one molecule of radical intermediate has, through its *p*-position, abstracted the nonaromatic H atom from a second molecule of radical intermediate to yield the normal end product of phenylation (biphenyl), itself thereby being converted into a dihydrobiphenyl. This overall reaction is known as **disproportionation**, because one molecule of radical intermediate has lost an H atom, while the other has gained one.

All these different potential end-products are commonly obtained, which means that radical aromatic substitution reactions—such as phenylation—often leave something to be desired as preparative methods. The nature of these several different products does, however, go a long way to establish that radical aromatic substitution does indeed proceed *via* intermediates of the kind suggested in [4.16] (p. 61).

4.2.3.2 Hydroxylation

It is also possible to introduce an OH group into the benzene ring of aromatic species, through reaction with hydroxyl radicals, $\cdot$OH. These are often generated *via* oxidation/reduction, e.g. with Fenton's reagent [4.4] (p. 53):

(1) $HO:OH + Fe^{2\oplus} \longrightarrow HO\cdot + {}^{\ominus}OH + Fe^{3\oplus}$

(2)

[4.22] HYDROXYLATION OF BENZENE

After formation of the usual type of delocalised intermediate, reaction of this with $Fe^{3\oplus}$ results in oxidative abstraction of the H atom as $H^{\oplus}$ to yield the product, phenol.

Like phenylation, hydroxylation of aromatic species is very rarely a preparative method of any significance; interest in it stems from its biological significance. This is due to the fact that the first step taken by animals, or indeed by human beings, to rid themselves of "foreign" aromatic molecules, that have been introduced into their systems, is hydroxylation of the benzene rings present in such molecules. Apart from any more specific metabolic reasons, hydroxylation greatly increases the solubility of aromatic compounds in water, thereby hastening their elimination from the organism.

4.3   SUMMARY

Initially there is a discussion of some of the methods by which **radicals** (species with an unpaired electron in their outer shell) may be generated, e.g. fission of an electron pair bond by <u>photolysis</u> (light) or <u>thermolysis</u> (heat), and single electron transfer *via* an <u>oxidation/reduction reaction</u>.

Many radical substitution reactions involve attack on C—H bonds which, because of their essentially non-polar character, are attacked with difficulty— if at all—by electrophiles and nucleophiles; a typical example of a radical substitution reaction—**halogenation**—is then discussed in detail. Overall substitution is here shown to involve <u>abstraction</u> of H· from R—H by an initiator radical, followed by attack on the halogenating agent, e.g. Cl_2, by the resultant alkyl radical, R·, to yield the product of overall substitution, R—Cl, and also Cl· which can "set off" an extremely rapid **chain reaction**. Consideration is then given to the influence of the structural environment on the reactivity towards H· abstraction of a C—H bond (primary, secondary and tertiary), to the effect of different halogenating agents, and to the effect of a double bond attached to the C—H bond carbon atom (in particular to the use of N-bromosuccinimide as a brominating agent).

Other radical substitution reactions considered are **autoxidation** (slow oxidative attack by oxygen at moderate temperatures), and its economic significance including its deleterious effects. Finally mention is made of radical **aromatic substitution**, particularly attack on aromatic species by $C_6H_5\cdot$ (<u>phenylation</u>) and by ·OH (<u>hydroxylation</u>), including the latter's biological significance.

Addition

5

Electrophilic addition

We have already seen (**1.6.2**, p.5) that the electron-rich nature of the carbon atoms in a carbon–carbon double bond means that they are most open to attack by reagents which are themselves electron-deficient, namely electrophiles; and that the nature of the resulting reaction is most likely to be addition.

5.1 ADDITION TO C=C

We mentioned, in (**1.6.2**, p. 5), the addition of bromine to such a double bond, and also the addition of two OH groups through its reaction with $KMnO_4$ (hydroxylation): both these addition reactions were long used as classical diagnostic tests for the presence of C=C.

5.1.1   Addition of bromine

We might well expect the addition of bromine to ethene to follow the simplest possible pathway, in which the two molecules just line up beside each other, and exchange electron pairs from existing bonds:

$$
\begin{array}{ccc}
\text{Br}-\text{Br} & & \text{Br}\quad\text{Br} \\
& \longrightarrow & \quad|\qquad| \\
\text{CH}_2=\text{CH}_2 & & \text{CH}_2-\text{CH}_2
\end{array}
$$

[5.1] EXCHANGE OF ELECTRON PAIRS BETWEEN Br$_2$ AND CH$_2$=CH$_2$

When we come to study the reaction a little more closely, however, we find two pieces of experimental evidence that throw considerable doubt on the validity of this simple, one step pathway.

The first of these involves what happens if the addition of bromine to ethene is carried out with a nucleophilic species—such as Cl$^\ominus$, $^\ominus$NO$_3$, H$_2$O:, etc.—also present in the solution: we then get not only the expected dibromide, but also a "mixed" addition product:

$$
\begin{array}{c}
\\
\\
\text{CH}_2=\text{CH}_2
\end{array}
\begin{array}{l}
\overset{\text{Br}_2/\text{Cl}^\ominus}{\nearrow} \quad
\begin{array}{cccc}
\text{Br} & \text{Br} & \text{Br} & \text{Cl} \\
| & | & | & | \\
\text{CH}_2 & \text{CH}_2 + & \text{CH}_2 & \text{CH}_2
\end{array} \\[2em]
\underset{\text{Br}_2/\text{NO}_3^\ominus}{\searrow} \quad
\begin{array}{cccc}
\text{Br} & \text{Br} & \text{Br} & \text{NO}_3 \\
| & | & | & | \\
\text{CH}_2 & \text{CH}_2 + & \text{CH}_2 & \text{CH}_2
\end{array}
\end{array}
$$

[5.2] "MIXED" ADDITION PRODUCT OBTAINED WHEN ADDED NUCLEOPHILE PRESENT

It is difficult to see how such "mixed" addition products could be formed if bromine addition was proceeding *via* the one-step pathway in [5.1] above. There is, however, the possibility that the "mixed" product is formed <u>after</u> initial bromine addition, by subsequent nucleophilic substitution on the first-formed dibromide by any nucleophile present in the solution:

$$
\begin{array}{ccc}
\begin{array}{cc}
\text{Br} & \text{Br} \\
| & | \\
\text{CH}_2- & \text{CH}_2 \\
\end{array}
& \longrightarrow &
\begin{array}{c}
\text{Br} \\
| \\
\text{CH}_2-\text{CH}_2 \\
| \\
\text{NO}_3
\end{array}
+ \text{Br}^\ominus
\end{array}
$$

[5.3] NUCLEOPHILIC SUSTITUTION ON THE FIRST-FORMED DIBROMIDE BY THE NUCLEOPHILE PRESENT IN SOLUTION

That this is **not** what is happening is confirmed, in a separate experiment, by reacting the nucleophile directly with the first-formed dibromide. It is then found that this substitution reaction, to form the "mixed" product, is considerably slower than the rate at which the "mixed" product is actually formed during the course of the original addition reaction. The "mixed" product must thus be produced by a route other than subsequent nucleophilic substitution of the first-formed dibromide.

The other piece of evidence involves steric considerations, but to investigate this point we need to consider addition to an alkene that has some substituents other than H on the carbon atoms of its double bond, e.g. MeCH=CHMe (but-2-ene). There are two quite different isomers of this compound, the *cis* and the *trans* forms, depending on whether the two Me groups in the molecule are on the <u>same</u> (*cis*), or <u>opposite</u> (*trans*), sides of the double bond:

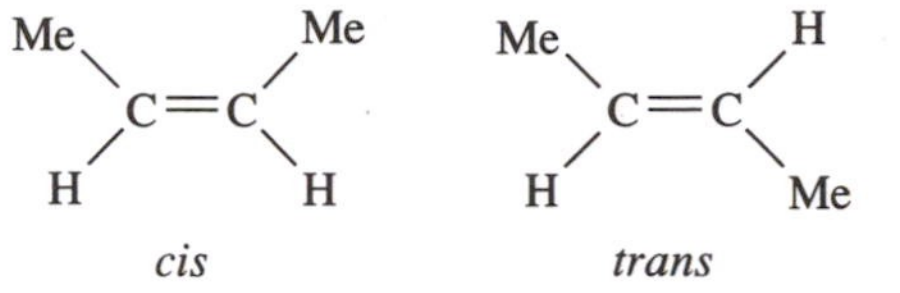

cis   *trans*

[5.4] *cis* AND *trans* BUT-2-ENES

If we consider addition of bromine to the *trans* isomer of but-2-ene, the simple one step pathway in [5.1] (p. 68) would require the addition of both bromine atoms to the same face of the planar (flat) alkene molecule:

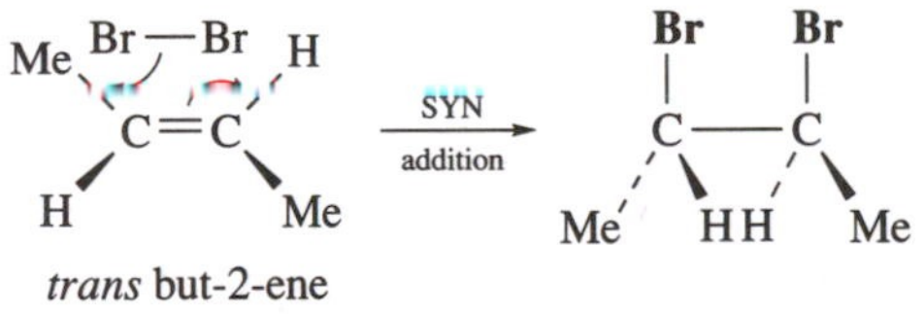

trans but-2-ene

[5.5] ADDITION OF BOTH BROMINE ATOMS TO THE SAME FACE OF *trans* BUT-2-ENE (SYN ADDITION)

Such same-face—called SYN—addition would lead to formation of the dibromide shown in [5.5], which has the two Me groups on <u>opposite</u> sides of the molecule. When we carry out the reaction in the laboratory, however, what we actually get is the isomeric dibromide which has the two Me groups attached to the <u>same</u> side of the molecule:

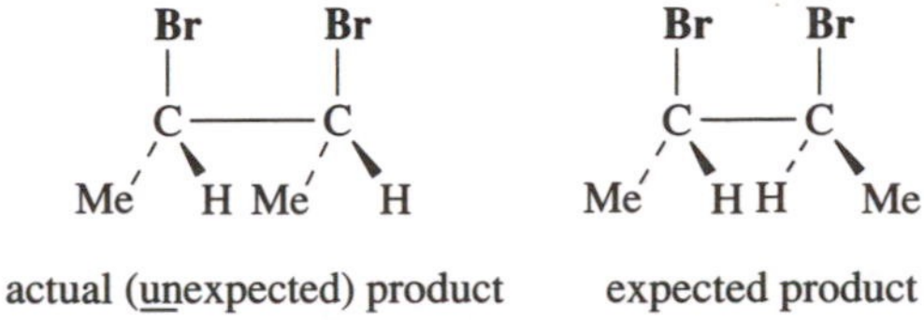

actual (<u>un</u>expected) product   expected product

[5.6] ACTUAL VERSUS EXPECTED PRODUCTS FROM ADDITION OF Br_2 TO *trans* BUT-2-ENE

This compound is clearly different from the dibromide we would have obtained if addition of Br_2 had occured *via* the SYN mode of [5.5] (p. 69); both these dibromides are known, and are readily distinguishable from each other.

To account for the formation of this unexpected product, the two bromine atoms must have added to opposite faces of the planar *trans* but-2-ene molecule:

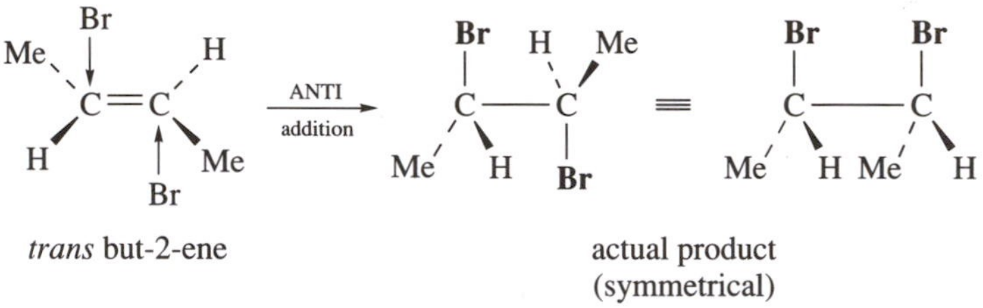

trans but-2-ene

actual product
(symmetrical)

[5.7] ADDITION OF THE TWO BROMINE ATOMS TO OPPOSITE
FACES OF *trans* BUT-2-ENE (ANTI ADDITION)

Such a mode of reaction is known as ANTI addition. There is no way in which the two Br atoms, in a molecule of bromine, can simultaneously add to the double bond of an alkene *via* the ANTI mode: the $\overline{Br—Br \; bond}$ just cannot stretch that far!

These two pieces of experimental evidence—the formation of mixed products, and ANTI addition—rule out a simple one step pathway for bromine addition ([5.1], p. 68), and suggest that reaction probably proceeds, therefore, *via* an intermediate. What is believed to happen is that the electron-rich alkene polarises one end of the bromine molecule (through repulsion of its electrons), thereby inducing a more electrophilic end in $^{\delta+}\underline{Br}—Br^{\delta-}$, and thus leading to bonding between bromine and the alkene:

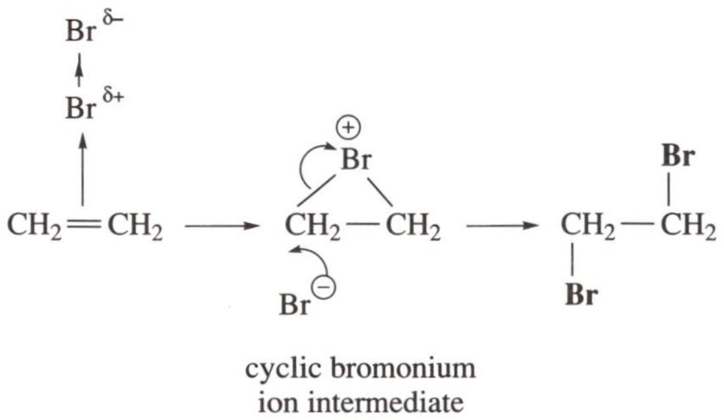

cyclic bromonium
ion intermediate

[5.8] ALKENE/Br_2 ADDITION *VIA* A CYCLIC BROMONIUM ION
INTERMEDIATE

This results in the formation of a **cyclic bromonium ion** intermediate, and attack on this cation by the nucleophilic $Br^\ominus$ (the residual half of the original Br_2 molecule) will lead to the end-product of overall addition, the dibromide. This nucleophilic attack by $Br^\ominus$ will take place from the side of the intermediate opposite to its $Br^\oplus$ atom, because this atom is very large, and will thus prevent access of $Br^\ominus$ to that side of the molecule.

The involvement of such an intermediate in bromine addition would also explain the other of our experimental observations. The formation of mixed products, when added nucleophiles are present, arises from competition between $Br^\ominus$ and the added nucleophile, $Y^\ominus$, in attack on the intermediate:

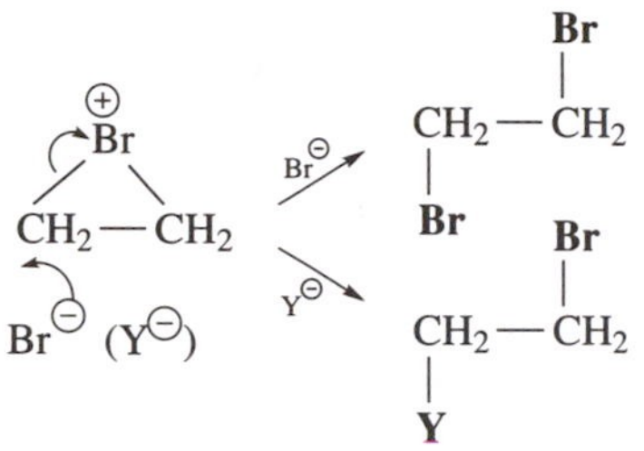

[5.9] COMPETITION BETWEEN $Br^\ominus$ AND $Y^\ominus$ IN ATTACK ON THE
INTERMEDIATE

It is, of course, also possible to explain mixed product formation on the basis of a simple carbocation as a rather less exotic intermediate:

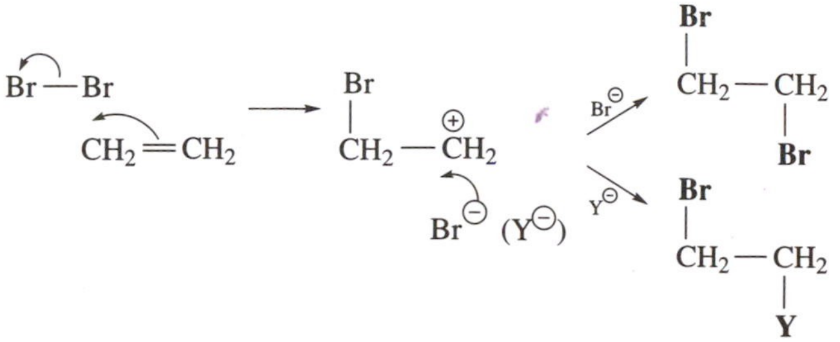

[5.10] MIXED PRODUCT FORMATION FROM A SIMPLE
CARBOCATION INTERMEDIATE

Such a carbocation intermediate <u>cannot</u>, however, account for the more precise stereochemical observation of overall ANTI addition, while a cyclic bromonium ion intermediate can:

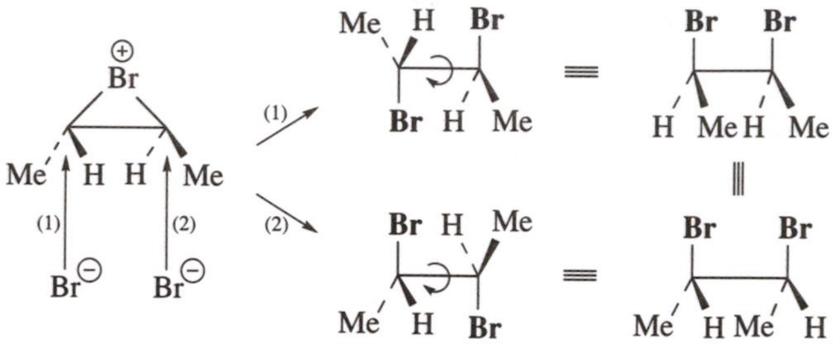

[5.11] OVERALL ANTI ADDITION OF Br$_2$ *VIA* A CYCLIC
BROMONIUM ION INTERMEDIATE

Attack on the intermediate by the residual Br$^\ominus$ (from the side of the molecule opposite to the bulky bromine atom) can be on either of the two carbon atoms of the original double bond. Attack on either does, however, lead to the same symmetrical dibromide: the known product of overall ANTI addition (*cf.* [5.6], p. 69).

5.1.1.1   Evidence for cyclic bromonium ion intermediates

The idea of cyclic bromonium ions as intermediates—admittedly of a somewhat unfamiliar type—has solved the problem of the otherwise anomalous experimental details of the addition of bromine; but it does raise the question of whether there is any independent evidence for their existence.

In fact the intermediate from the addition of bromine to the alkene, Me$_2$C=CMe$_2$, has been identified, and its structure confirmed, spectro-

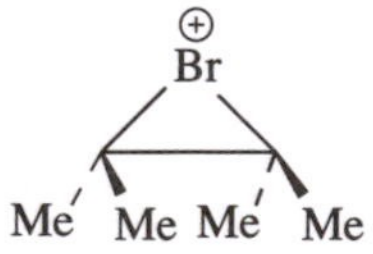

[5.12] CYCLIC BROMONIUM ION INTERMEDIATE FROM ADDITION
OF Br$_2$ TO Me$_2$C=CMe$_2$

scopically. It has also proved possible actually to isolate the intermediate obtained in the addition of Br$_2$ to the rather unusual alkene in [5.13]:

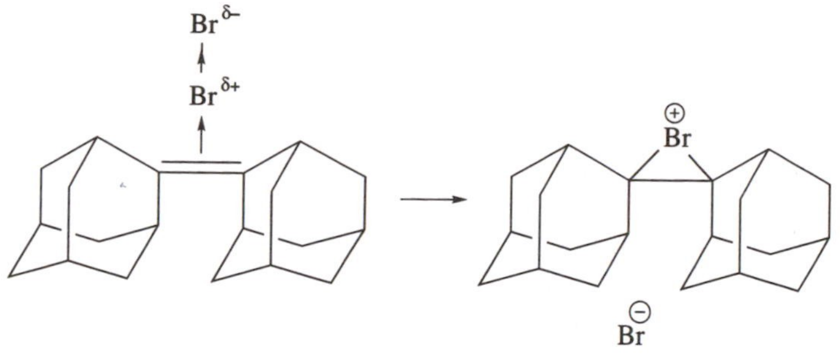

[5.13] ISOLATION OF A CYCLIC BROMONIUM ION INTERMEDIATE

The possibility of actually isolating the intermediate in this case is due, no doubt, to the extremely bulky cage-like structures attached to each of the two carbon atoms of the original double bond; these are large enough entirely to prevent access by $Br^{\ominus}$ to complete the overall addition of Br_2. The existence of cyclic bromonium ions, involving a three-membered ring, seems less unusual when we realise that similar three-membered rings, epoxides (**5.1.4.2**, p. 78)—involving another electronegative atom, oxygen— are both common, and stable.

5.1.2 Addition of other halogens

We have to date concentrated on the addition of bromine, and reference must now be made to the other halogens. Direct reaction with fluorine, as with substitution, is usually so vigorous that C—C bond-breaking may occur: it is thus of no preparative value. Addition of chlorine proceeds readily, but because Cl is more electronegative than Br it is somewhat less prepared to share an electron pair with an adjacent cationic carbon atom:

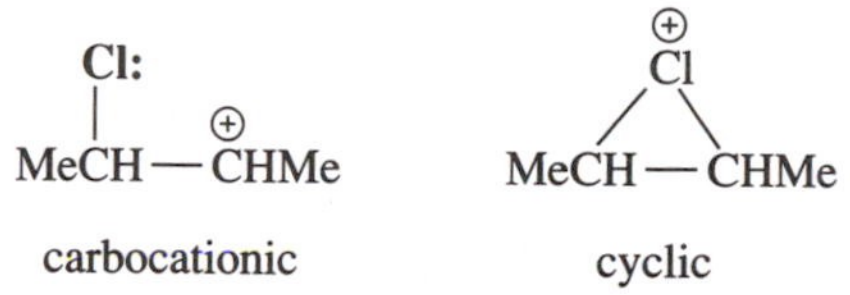

carbocationic   cyclic

[5.14] CARBOCATIONIC *VERSUS* CYCLIC INTERMEDIATES IN ADDITION OF Cl_2

There is thus less tendency to form cyclic intermediates, and simple carbocations are often involved. This is reflected in the observation that the addition of Cl_2 is often found to proceed by both ANTI and SYN modes simultaneously, whereas reaction entirely *via* a cyclic chloronium ion intermediate would have resulted in ANTI addition only.

Addition of both chlorine and bromine is accelerated by the prescence of Lewis acids, e.g. $AlCl_3$, $FeBr_3$, because they are able to polarise the halogen molecule, thus inducing in it a more electrophilic—and hence more reactive—end (*cf.* **3.2.2**, p. 36). Direct addition of iodine is also of little preparative value, largely because of the ease with which the reaction may be reversed.

5.1.2.1 Effect of structure of alkene

Irrespective of whether the intermediate involved in halogen addition is a cyclic halonium ion or a simple carbocation, we might expect any substituent—on the carbon atoms of the original double bond—which is electron-donating to facilitate addition, through stabilisation of a developing intermediate which carries a positive charge. This is borne out by the relative rates of addition of bromine to the alkenes shown in [5.15]:

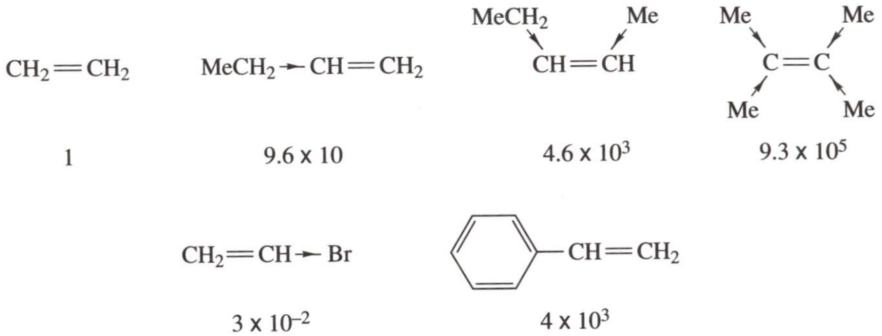

$$CH_2\!=\!CH_2 \qquad MeCH_2\!\rightarrow\!CH\!=\!CH_2 \qquad \begin{matrix} MeCH_2 \\ \diagdown \\ CH \end{matrix}\!=\!\begin{matrix} Me \\ \diagup \\ CH \end{matrix} \qquad \begin{matrix} Me \\ \diagdown \\ Me \end{matrix}\!\!\begin{matrix} \diagup \\ C \\ \diagup \end{matrix}\!=\!\begin{matrix} \diagdown \\ C \\ \diagdown \end{matrix}\!\!\begin{matrix} Me \\ \\ Me \end{matrix}$$

1   9.6 x 10   4.6 x 10^3 9.3 x 10^5

$$CH_2\!=\!CH\!\rightarrow\!Br \qquad\qquad \text{(benzene ring)}\!-\!CH\!=\!CH_2$$

3 x 10^{-2} 4 x 10^3

[5.15] EFFECT OF SUBSTITUENTS ON RELATIVE RATES OF ADDITION OF Br$_2$ TO C=C

It will also be observed that the electron-withdrawing substituent, Br, slows addition down.

A benzene ring attached to a double bond carbon atom is also found to facilitate addition. In this case the cationic intermediate is more likely to be the simple carbocation, whose developing +^{ve} charge can be delocalised over the carbon atoms of the benzene ring (*cf.* **3.2.5.1.1**, p. 42), thus stabilising it and making its formation easier:

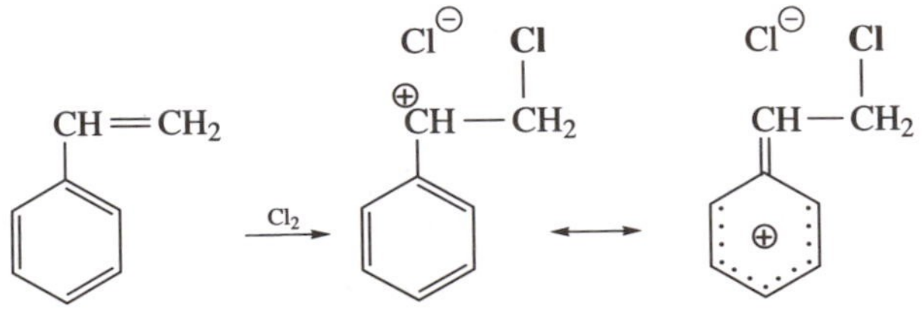

[5.16] STABILISATION OF A CARBOCATION INTERMEDIATE BY A BENZENE RING

5.1.3   Addition of HX

All the hydrogen halides, HX (X = F, Cl, Br, I), will add to C=C, their relative ease of doing so following the order of their strength as acids— HI > HBr > HCl > HF. This suggests that protonation of the double bond is likely to be the controlling step of the overall reaction:

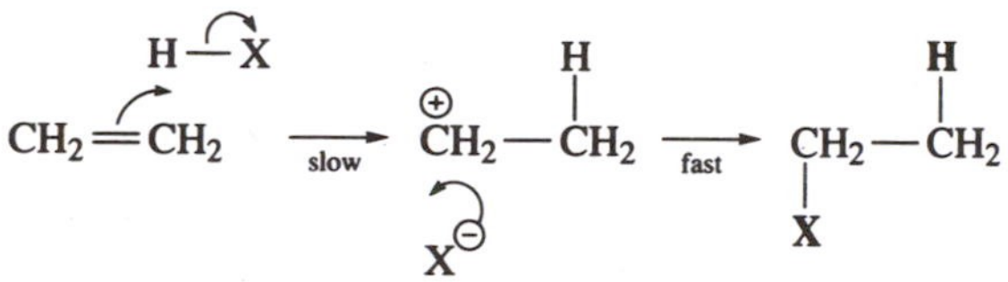

$$CH_2=CH_2 \xrightarrow[\text{slow}]{H-X} \overset{\oplus}{CH_2}-CH_2 \quad X^{\ominus} \xrightarrow{\text{fast}} \overset{H}{\underset{}{CH_2}}-\overset{H}{\underset{X}{CH_2}}$$

[5.17] ADDITION OF HX TO C=C

A simple carbocation will be the intermediate, as H has no electron pair to share with the adjacent cationic centre.

When HX is added to an unsymmetrical alkene, such as Me—CH=CH$_2$, then a new problem arises—that of **orientation** of addition.

5.1.3.1   Orientation of addition

In the addition of HBr to Me—CH=CH$_2$, initial protonation of the double bond could, in theory, take place at either of its two carbon atoms leading to different carbocation intermediates:

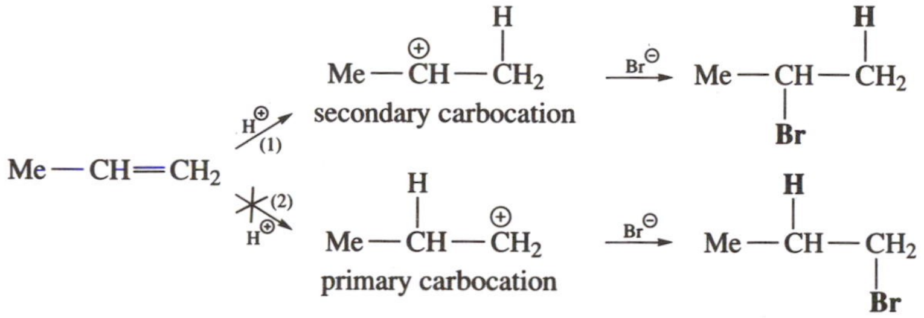

Me—CH=CH$_2$

(1) $\overset{H}{Me-\overset{\oplus}{CH}-CH_2}$ secondary carbocation $\xrightarrow{Br^{\ominus}}$ $\overset{H}{Me-CH-\underset{Br}{CH_2}}$

(2) $\overset{H}{Me-CH-\overset{\oplus}{CH_2}}$ primary carbocation $\xrightarrow{Br^{\ominus}}$ $\overset{H}{Me-CH-\underset{Br}{CH_2}}$

[5.18] ORIENTATION: ADDITION OF HBr TO Me—CH=CH$_2$

Reaction of these intermediates with the residual Br$^{\ominus}$ from HBr would result in the formation of different products from overall addition. The product that is actually obtained is the one derived *via* pathway (1); this happens because the relative stability of alkyl carbocations is known to follow the sequence shown in [5.19]:

$$
\underset{\text{tertiary}}{\overset{\displaystyle R}{\underset{\displaystyle R}{R \to \overset{\oplus}{C}}}} \quad > \quad \underset{\text{secondary}}{\overset{\displaystyle R}{\underset{\displaystyle R}{\overset{\oplus}{CH}}}} \quad > \quad \underset{\text{primary}}{R \to \overset{\oplus}{CH_2}}
$$

[5.19] RELATIVE STABILISATION OF CARBOCATIONS

The secondary carbocation, in pathway (1), is more stable than the primary carbocation, in pathway (2), and is therefore formed preferentially; the addition product actually obtained is thus, Me—CH(Br)—CH$_3$ (2-bromopropane) derived *via* pathway (1).

This is known as **Markownikov** addition, so named after the inventor of an empirical rule, which can be expressed as: "In the addition of an unsymmetrical reagent to an unsymmetrical alkene, under polar conditions, the more negative moiety of the reagent becomes attached to the more highly substituted of the two carbon atoms of the original double bond". This empirical rule, derived from experimental observations, is justified theoretically on the basis of the relative stabilisation of carbocations, as we have already seen ([5.19]). When addition takes place under non-polar (radical) conditions (**6.1.2**, p. 88), a different pattern of orientation is observed.

5.1.3.2 Hydration

Acids that ionise to produce weakly nucleophilic anions, e.g. $HSO_4^{\ominus}$ from H_2SO_4, may be used, in dilute aqueous solution, to add—overall—H_2O to alkenes: this is known as **hydration**. Protonation of the alkene first takes place and, if the acid anion is weak, the resultant carbocation intermediate is then attacked preferentially by the reasonably good nucleophile, H_2O:, which is present in large excess:

$$
Me_2C{=}CH_2 \xrightleftharpoons{H_2SO_4} \underset{H_2O:}{Me_2\overset{\oplus}{C}-CH_2} \overset{H}{\underset{H_2O:}{\xrightleftharpoons{}}} \underset{H_2\overset{\oplus}{O}}{Me_2\overset{\displaystyle H}{C}-CH_2} \xrightleftharpoons{} \underset{HO}{Me_2\overset{\displaystyle H}{C}-CH_2}
$$

[5.20] ACID-CATALYSED HYDRATION OF ALKENES

Finally the adduct loses $H^{\oplus}$, leading overall to hydration of C=C.

If any attack by the acid anion, e.g. $HSO_4^{\ominus}$, on the carbocation intermediate does take place, then the added anion is found to undergo extremely ready nucleophilic displacement by the high concentration of H_2O: in the solution. The overall reaction can thus be considered, essentially, as the acid-catalysed addition of H_2O to alkenes. The reaction is readily reversible and, in the reverse direction, is the well-known acid-catalysed dehydration of alcohols (**9.1.1**, p. 142).

5.1.4 Hydroxylation

We have already mentioned (**1.6.2**, p. 5) the reaction of alkenes with $KMnO_4$ as a classical diagnostic test for unsaturation, and seen that it results in the overall addition of two OH groups to the carbon–carbon double bond: this is known as **hydroxylation**.

5.1.4.1 SYN hydroxylation

It is found that this reaction with $KMnO_4$ results in addition proceeding entirely *via* the SYN mode. This specificity is believed to result from the reaction involving the formation of a cyclic permanganic ester as an intermediate, and its subsequent hydrolysis, e.g. in the hydroxylation of *cis* but-2-ene:

cis but-2-ene   intermediate

[5.21] HYDROXYLATION OF *cis* BUT-2-ENE *VIA* A CYCLIC PERMANGANIC ESTER INTERMEDIATE

The cyclic ester intermediate is readily hydrolysed in the aqueous solution to yield the 1,2-diol: the product of overall SYN addition.

The major drawback to the use of $KMnO_4$ for preparative hydroxylation is that this reagent is a powerful non-specific oxidising agent, which readily oxidises the initial 1,2-diol further. This can be prevented to some extent by carrying out the reaction under alkaline conditions ($KMnO_4$ is then less powerful as an oxidising agent), and in dilute solution. It has not proved possible actually to isolate cyclic permanganic ester intermediates, but they have been detected spectroscopically.

Hydroxylation may also be carried out using the very similar reagent osmium tetroxide, OsO_4. This too leads, hardly surprisingly, to specific SYN addition of two OH groups to *cis* but-2-ene; but this time it <u>is</u> possible to isolate, and identify, the cyclic osmic ester intermediate:

cis but-2-ene   intermediate

[5.22] HYDROXYLATION OF *cis* BUT-2-ENE *VIA* A CYCLIC OSMIC ESTER INTERMEDIATE

As in the $KMnO_4$ reaction, hydrolysis of the cyclic intermediate then yields the product 1,2-diol. Osmium tetroxide is both expensive and highly toxic, and is therefore commonly used preparatively in only catalytic amounts, which can be accomplished by carrying out the reaction in the presence of hydrogen peroxide, H_2O_2. The hydrogen peroxide oxidises the H_2OsO_4, produced from hydrolysis of the cyclic osmic ester, back to OsO_4 again, which is then able to add to a further molecule of alkene, and so the cycle goes on.

5.1.4.2   ANTI hydroxylation

It is also possible to arrange for hydroxylation of C=C to proceed entirely *via* the ANTI mode: in this case the intermediate is quite stable, and may readily be isolated. Attack on a double bond by a peroxy acid—such as peroxybenzoic acid, $C_6H_5CO_3H$—leads to addition of an oxygen atom across the double bond, e.g. in *trans* but-2-ene, to form a stable three-membered ring, an **epoxide**:

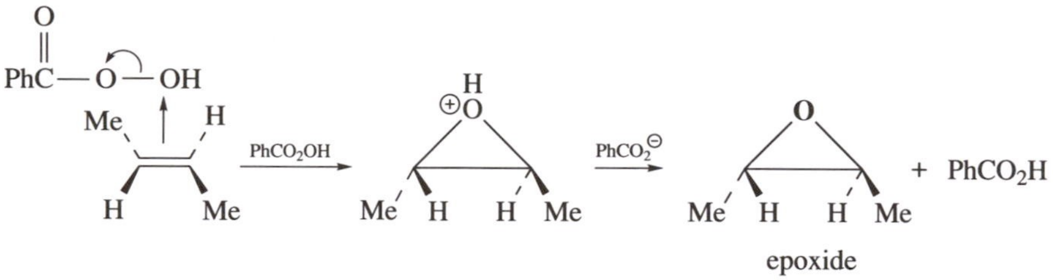

[5.23] FORMATION OF AN EPOXIDE FROM *trans* BUT-2-ENE

The epoxide closely resembles a cyclic bromonium ion intermediate (*cf.* [5.12], p. 72), though it is much more stable. It readily undergoes attack by nucleophiles, e.g.$^\ominus$OH, which can—as with cyclic bromonium ions—take place on either carbon atom of the original double bond from the side of the epoxide molecule opposite to its oxygen atom:

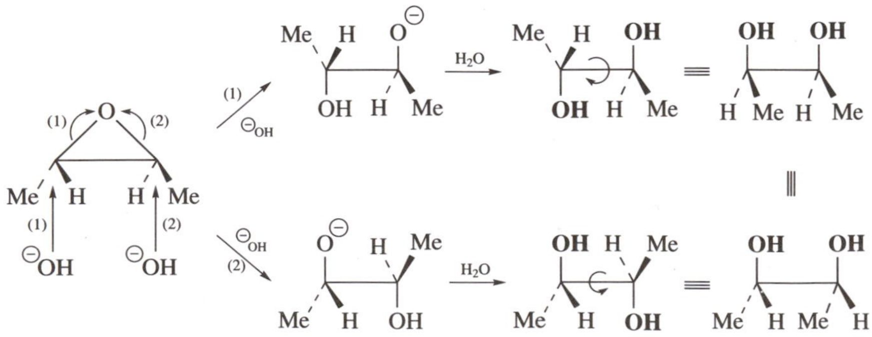

[5.24] BASE HYDROLYSIS OF EPOXIDE FROM *trans* BUT-2-ENE

Attack on either carbon atom leads to the same 1,2-diol: the product of overall ANTI addition to *trans* but-2-ene.

Hydrolysis of an epoxide can also take place under acidic conditions, the first step being protonation of its oxygen atom. The positively charged epoxide can now be attacked by considerably weaker nucleophiles, e.g. H_2O:

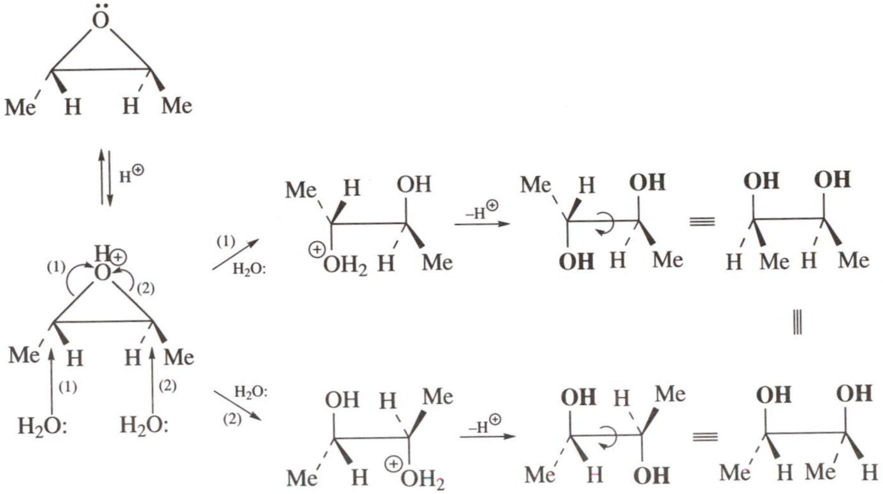

[5.25] ACID HYDROLYSIS OF EPOXIDE FROM *trans* BUT-2-ENE

Again, attack on either carbon atom leads to the same 1,2-diol: the product of overall ANTI addition to *trans* but-2-ene.

5.1.5 Cationic polymerisation

We have already seen ([5.17], p. 75) that addition of $H^\oplus$ converts a carbon–carbon double bond into a carbocation. Such an electron-deficient species must be capable of acting as an electrophile, and the point then arises as to whether carbocations are themselves thus able to add to C=C. We find in practice that provided there are no powerful nucleophiles present, that would "mop up" carbocations as soon as they were formed, then addition to C=C does indeed occur.

Thus with 2-methylpropene, $Me_2C=CH_2$ ([5.26], p. 80), initial protonation takes place on the double bond carbon atom that will lead to the more stable of the two possible carbocations (tertiary rather than primary). This carbocation can then add to the double bond of another molecule of $Me_2C=CH_2$ (again so as to give the more stable of the two possible carbocations) to yield a new, longer carbocation, which in turn can add to yet another molecule of $Me_2C=CH_2$, and so it goes on. Under suitable conditions, long chain polymer molecules may be produced (e.g. butyl rubber from $Me_2C=CH_2$), but in general radical-induced polymerisation of alkenes (**6.1.3**, p. 90) is of greater importance.

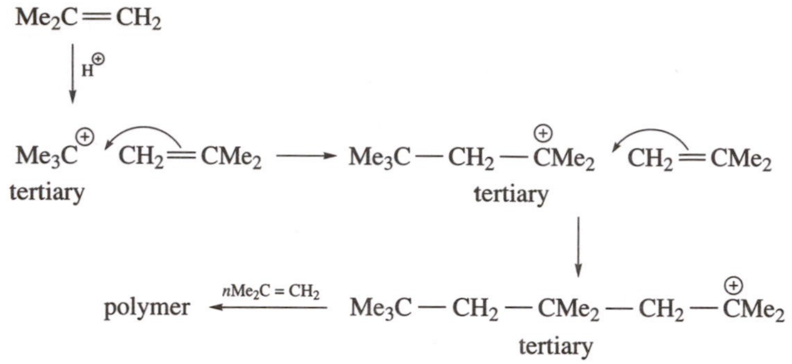

$$Me_2C{=}CH_2$$

$$\downarrow H^{\oplus}$$

$$Me_3\overset{\oplus}{C} \quad CH_2{=}CMe_2 \longrightarrow Me_3C{-}CH_2{-}\overset{\oplus}{C}Me_2 \quad CH_2{=}CMe_2$$

tertiary   tertiary

$$\downarrow$$

polymer $\xleftarrow{\; n Me_2C\,=\,CH_2 \;}$ $Me_3C{-}CH_2{-}CMe_2{-}CH_2{-}\overset{\oplus}{C}Me_2$

tertiary

[5.26] CATIONIC POLYMERISATION OF 2-METHYLPROPENE

5.2 ADDITION TO C=C—C=C

Where carbon–carbon double bonds alternate in a compound with carbon–carbon single bonds, they are said to be **conjugated**. The occurrence of conjugation in a compound is found to influence the reactivity of such double bonds towards electrophiles, compared with double bonds that are not conjugated (**isolated** double bonds). Thus addition of electrophiles, such as Br_2 and HBr, is found to proceed rather more rapidly with $CH_2{=}CH{-}CH{=}CH_2$ (buta-1,3-diene) than it does with $CH_2{=}CH_2$, or $CH_2{=}CH{-}CH_2{-}CH_3$.

5.2.1 Addition of bromine

The reason for this somewhat greater ease of reaction is that the carbocation intermediate, formed from initial electrophilic attack by Br_2 on buta-1,3-diene, is stabilised—compared with the intermediate from a similar attack on ethene—through delocalisation of its charge (*cf.* [2.15], p. 19), and is therefore formed more readily:

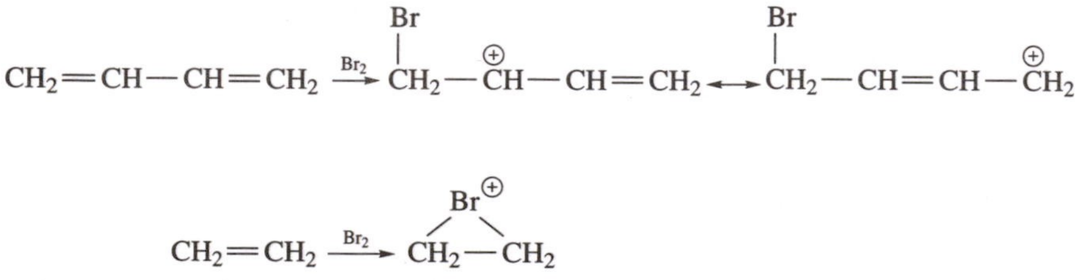

$$CH_2{=}CH{-}CH{=}CH_2 \xrightarrow{Br_2} \overset{\overset{\textstyle Br}{|}}{CH_2}{-}\overset{\oplus}{CH}{-}CH{=}CH_2 \longleftrightarrow \overset{\overset{\textstyle Br}{|}}{CH_2}{-}CH{=}CH{-}\overset{\oplus}{CH_2}$$

$$CH_2{=}CH_2 \xrightarrow{Br_2} \overset{\overset{\textstyle \overset{\oplus}{Br}}{\diagup\diagdown}}{CH_2{-}CH_2}$$

[5.27] COMPARISON OF INTERMEDIATES FROM ATTACK OF Br_2
ON BUTA-1,3-DIENE AND ETHENE

Attack on buta-1,3-diene can take place on either of the two terminal carbon atoms to form the same, secondary carbocation intermediate; initial attack on either of the non-terminal carbon atoms would lead to the formation of the less stable primary carbocation. A simple carbocation intermediate is formed, rather than a cyclic bromonium ion as with ethene, because a carbocation allows stabilisation through delocalisation of its $+^{ve}$ charge, whereas a cyclic bromonium ion would not.

The occurrence of delocalisation in the carbocation intermediate means that completion of addition, by attack of $Br^{\ominus}$, can take place at two different positions leading overall to <u>either</u> 1,2- <u>or</u> 1,4-addition (or to a mixture of both):

[5.28] COMPLETION OF OVERALL 1,2- AND/OR 1,4-ADDITION OF Br_2 TO BUTA-1,3-DIENE

Both modes of addition tend to occur, and a mixture of addition products is commonly obtained. The composition of this mixture is influenced by the conditions, higher temperatures tending to favour 1,4-addition; this latter mode is sometimes referred to as <u>conjugate</u> addition.

5.2.2 Addition of HBr

As with simple alkenes, addition of an unsymmetrical adduct to an unsymmetrical conjugated diene raises the question of orientation of addition. In the light of our previous experience (*cf.* [5.18], p. 75) however, this poses no new problems.

Thus in the addition of HBr to $MeCH{=}CH{-}CH{=}CH_2$ ([5.29], p. 82), initial protonation could take place on either of the two carbon atoms at the ends of the conjugated system (protonation on either of the two internal carbon atoms would lead to less stable carbocations, *cf.* **5.2.1**, p. 80). Two alternative delocalised carbocations could thus, in theory, be obtained, but the one at the top in [5.29] has a contributing canonical structure (underlined) which is a primary cation, and the delocalised intermediate to which it contributes will thus be less stabilised than the one below, which has only contributing canonical structures which are secondary cations. Only the

$$\underset{\substack{|\\ \text{MeCH}-\overset{\oplus}{\text{CH}}-\text{CH}=\text{CH}_2}}{\overset{\text{H}}{}} \longleftrightarrow \underset{\substack{|\\ \text{MeCH}-\text{CH}=\text{CH}-\overset{\oplus}{\text{CH}}_2}}{\overset{\text{H}}{}}$$

$$\cancel{\;\;} \uparrow \text{H}^{\ominus}$$

$$\text{MeCH}=\text{CH}-\text{CH}=\text{CH}_2$$

$$\downarrow \text{H}^{\ominus}$$

$$\underset{\substack{|\\ \text{MeCH}=\text{CH}-\overset{\oplus}{\text{CH}}-\text{CH}_2}}{\overset{\text{H}}{}} \longleftrightarrow \underset{\substack{|\\ \overset{\oplus}{\text{MeCH}}-\text{CH}=\text{CH}-\text{CH}_2}}{\overset{\text{H}}{}}$$

$$\downarrow \text{Br}^{\ominus} \qquad\qquad\qquad \downarrow \text{Br}^{\ominus}$$

$$\underset{\substack{|\\ \text{Br}}}{\underset{\substack{|\\ \text{MeCH}=\text{CH}-\text{CH}-\text{CH}_2}}{\overset{\text{H}}{}}} \qquad \underset{\substack{|\\ \text{Br}}}{\underset{\substack{|\\ \text{MeCH}-\text{CH}=\text{CH}-\text{CH}_2}}{\overset{\text{H}}{}}}$$

1,2 - addition   1,4 - addition

[5.29] ADDITION OF HBr TO MeCH=CH—CH=CH$_2$

lower intermediate is thus formed, which can, of course, then undergo attack by Br$^{\ominus}$ at either of the two cationic centres in the delocalised intermediate leading overall to 1,2- or 1,4-addition of HBr (or, of course, to a mixture of both possible products).

5.3   ADDITION TO C≡C

Hardly surprisingly, electrophiles are also found to add to alkynes, whose carbon–carbon triple bonds should be even more electron-rich than the double bonds in alkenes whose reactions we have been considering. It is, therefore, somewhat of a surprise to find that the reactions of alkynes with electrophiles are usually slower than those of analogous alkenes, under comparable conditions.

This may be due in small part to the fact that the carbon atoms in a triple bond are found to be considerably more electronegative than the ones in a double bond; they can thus keep a somewhat tighter grip on the electrons in the bond, and are thus correspondingly less willing to share them with an attacking electrophile. This greater hold on the electrons, by the carbon atoms in a triple bond, is reflected in the fact that ethyne (HC≡CH) is found to be markedly acidic: on reaction with strong bases, e.g. $^{\ominus}$NH$_2$' it loses a proton to form the stable anion, HC≡C$^{\ominus}$, behaviour quite different from that of ethene.

The major reason for the lower reactivity of alkynes, CH≡CR, towards electrophiles is, however, that the vinyl cations, e.g. CH$_2$=CR$^{\oplus}$, formed as intermediates on initial addition of electrophiles, e.g. $H^\oplus$, are markedly less stable than the alkyl cations, e.g. $CH_3 — CHR^\oplus$, formed on similar addition to alkenes, $CH_2{=}CHR$, and are formed correspondingly less readily.

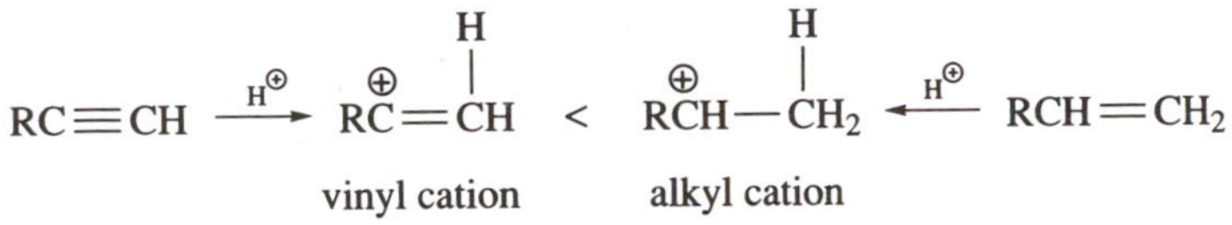

$$RC{\equiv}CH \xrightarrow{H^\oplus} \underset{\text{vinyl cation}}{R\overset{\oplus}{C}{=}\overset{H}{C}H} \quad < \quad \underset{\text{alkyl cation}}{R\overset{\oplus}{C}H—\overset{H}{C}H_2} \xleftarrow{H^\oplus} RCH{=}CH_2$$

[5.30] RELATIVE STABILITY OF CATIONIC INTERMEDIATES FROM PROTONATION OF ALKYNES AND ALKENES

When an electrophile, such as HBr, adds to a triple bond as in [5.31], the reaction may often be stopped at the half-way stage, after only one molecule of electrophile has added on. The orientation of addition of the second molecule of HBr is governed by Markownikov's rule (**5.1.3.1**, p. 76), but this second addition will be slower than the first because of the electron-withdrawing effect of the Br atom (*cf.* [5.15], p. 74); hence the possibility of stopping the reaction half-way, if so desired:

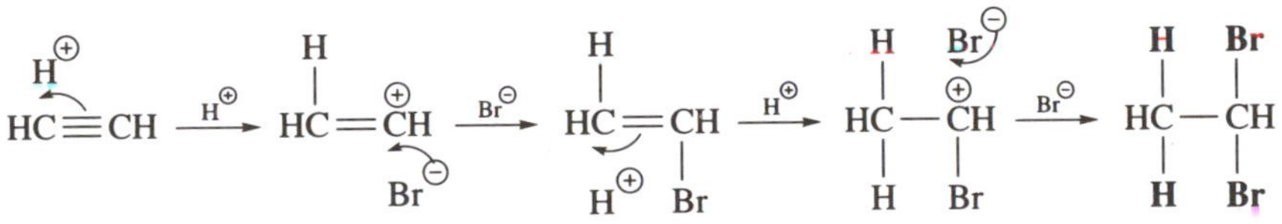

$$HC{\equiv}CH \xrightarrow{H^\oplus} \underset{Br^\ominus}{HC{=}\overset{\oplus}{C}H} \xrightarrow{Br^\ominus} \underset{H^\oplus\;Br}{HC{=}CH} \xrightarrow{H^\oplus} \underset{H\;Br}{HC{-}\overset{\oplus}{C}H} \xrightarrow{Br^\ominus} \underset{H\;Br}{HC{-}CH}$$

[5.31] ADDITION OF HBr TO HC≡CH

Alkynes may also be hydrated by reacting them with aqueous solutions of strong acids, and Markownikov's rule again applies:

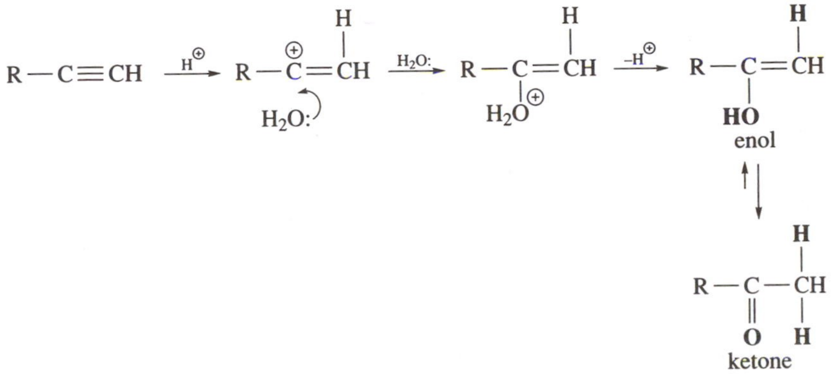

$$R{-}C{\equiv}CH \xrightarrow{H^\oplus} \underset{H_2O:}{R{-}\overset{\oplus}{C}{=}\overset{H}{C}H} \xrightarrow{H_2O:} \underset{H_2\overset{\oplus}{O}}{R{-}C{=}\overset{H}{C}H} \xrightarrow{-H^\oplus} \underset{\substack{HO\\ \text{enol}}}{R{-}C{=}\overset{H}{C}H}$$

$$\updownarrow$$

$$\underset{\substack{O\;\;H\\ \text{ketone}}}{R{-}\overset{\text{H}}{C}{-}CH}$$

[5.32] HYDRATION OF ALKYNES

The product is the unstable **enol** form of a ketone, which rapidly reverts to the normal **keto** form; this reaction can be a useful synthetic route to ketones.

5.4 ADDITION TO C=O

In the carbonyl group, C=O, the electrons in the carbon–oxygen double bond are not shared equally between the two atoms; this results from oxygen being more electronegative than carbon, and so drawing the electrons of the double bond towards itself, and away from carbon: the C=O bond is thus polarised (*cf.* C—Br in [1.8], p. 4):

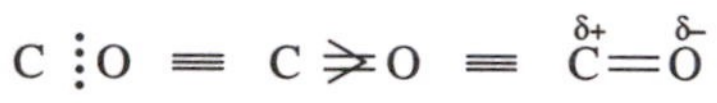

$$C \ \vdots \ O \ = \ C \gg O \ = \ \overset{\delta+}{C} = \overset{\delta-}{O}$$

[5.33] POLARISATION OF C=O

We could thus imagine overall addition to C=O being initiated either by electrophilic attack on oxygen or by nucleophilic attack on carbon:

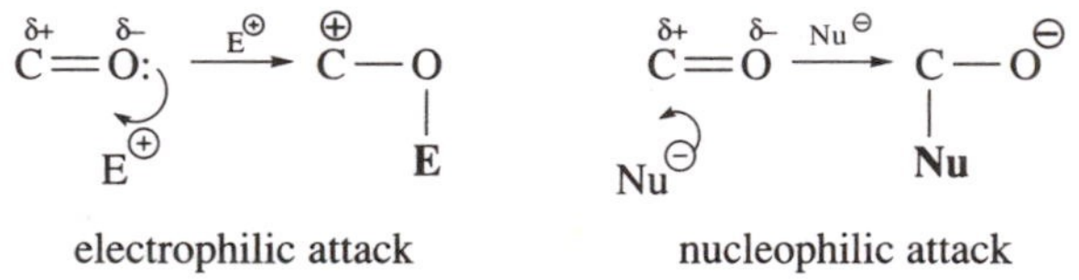

electrophilic attack   nucleophilic attack

[5.34] ALTERNATIVE INITIAL ATTACKS ON C=O

In fact, the initial electrophilic attack on oxygen is of real significance only as acid catalysis of the overall addition of nucleophiles, such as ROH, which would otherwise be too weakly nucleophilic to react with C=O:

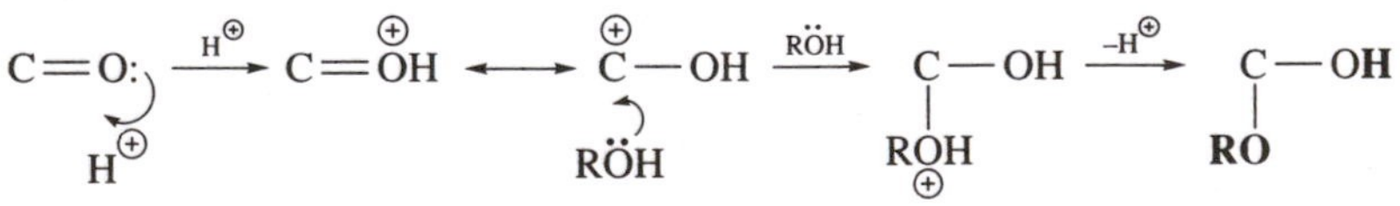

[5.35] ACID-CATALYSIS OF THE ADDITION OF WEAK NUCLEOPHILES

Initial protonation of the oxygen atom produces a $+^{ve}$ charge on the carbon atom which is then attacked much more readily, even by weak nucleophiles. The major pattern of addition to C=O is, however, by nucleophiles (**7.2**, p. 105).

5.5 SUMMARY

The electron-rich nature of double bonds suggests that they will be attacked most readily by electron-deficient reagents—**electrophiles**—and that the overall reaction will be **addition**.

The addition of bromine is discussed first, and evidence is presented that this does not involve simple exchange of electron pairs between C=C and Br—Br, but proceeds *via* a **cyclic bromonium ion** intermediate. The reaction pathway for addition that is followed by the other halogens is also discussed.

The addition of hydrogen halide, HX, is then considered and how, with non-symmetrical alkenes, this raises the problem of **orientation** of addition. Orientation of addition can be predicted by the empirical generalisation, **Markownikov's rule**, for which a rational explanation is provided in terms of the relative stability of the potential intermediate (**carbocation**). Other addition reactions are then considered, e.g. acid-catalysed hydration, hydroxylation (*via* both SYN and ANTI modes), and cationic polymerisation.

Addition of electrophiles such as Br_2 and HBr to conjugated systems, e.g. C=C—C=C, is then considered, including the possible formation of either 1,2- or 1,4-addition products, or of a mixture of both. The problem of orientation, when HBr is added to a non-symmetrical system, is also discussed.

Addition of electrophiles to C≡C is found, slightly surprisingly, to be slower than similar addition to comparable C=C; this is explainable in terms of relative intermediate stability. Addition to C≡C can, in some cases, be stopped after only one molecule of electrophile has reacted, e.g. the addition of HBr, and the acid-catalysed addition of H_2O, which affords a useful synthetic route to ketones.

Finally, reference is made to electrophilic addition to C=O, which would involve initial attack by the electrophile on oxygen. In fact, such attack is all but confined to protonation, in order to assist nucleophilic attack at carbon.

6

Radical addition

The great majority of radicals will add readily to carbon–carbon double bonds; this includes radicals derived from species such as Br_2 and HBr, which we have already seen (**5.1.1**, p. 68 and **5.1.3**, p. 75) acting as electrophiles towards C=C. In general it is found that polar solvents, and Lewis acid catalysts, promote electrophilic addition of such species, while radical addition is promoted by light, the presence of introduced initiator radicals, non-polar solvents, or by carrying out the reaction in the gas phase. Perhaps the most important radical addition reaction, and certainly the one that has been most intensively studied, is the large scale production of polymers from alkenes: vinyl polymerisation (**6.1.3**, p. 90).

6.1 ADDITION TO C=C

Among the most straightforward radical addition reactions is that of the halogens.

6.1.1 Addition of halogens

The relative reactivity of the halogens in adding to C=C *via* a radical pathway is found to follow the order: $F_2 > Cl_2 > Br_2 > I_2$; the same order as was observed for addition *via* an electrophilic pathway. The reaction with fluorine needs no initiation, and proceeds so vigorously as to make it of no preparative value. The addition of chlorine may be initiated photochemically (*cf.* **4.2.1**, p. 54) to yield Cl·, and a chain reaction is then set up as shown in [6.1]:

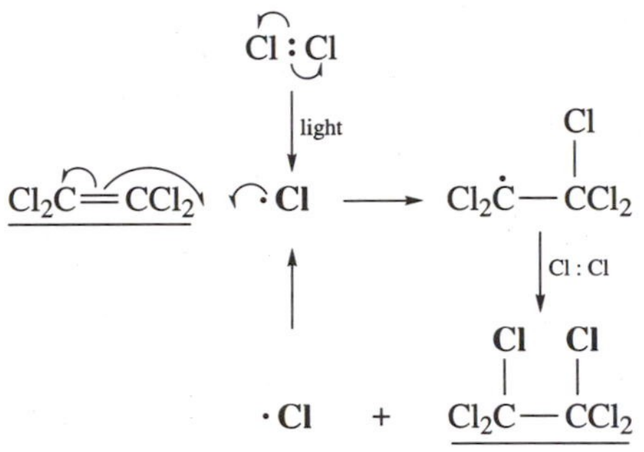

$$\text{Cl : Cl}$$

$$\Big\downarrow \text{light}$$

$$Cl_2C{=}CCl_2 \quad \cdot Cl \longrightarrow Cl_2\dot{C}{-}CCl_2$$

$$\Big\downarrow Cl : Cl$$

$$\cdot Cl \quad + \quad Cl_2C{-}CCl_2$$

[6.1] CHAIN REACTION IN ADDITION OF Cl_2 TO C=C

Each input of a quantum of light produces <u>two</u> chlorine radicals, <u>each</u> of which initiates the conversion of several thousand molecules of unsaturated starting material into addition product. These chain reactions come to a stop only when the concentration of starting materials has sunk to a pretty low level. Only then does the concentration of radicals become sufficiently large—relative to that of starting materials—for there to be a significant chance of two radicals reacting with each other, rather than with further molecules of $Cl_2C{=}CCl_2$ or Cl_2, (*cf.* **4.2.1**, p. 54):

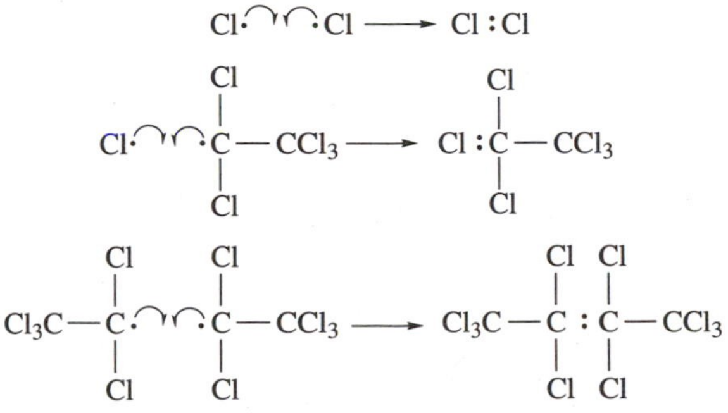

$$Cl\cdot \frown \cdot Cl \longrightarrow Cl : Cl$$

$$Cl\cdot \frown \cdot C{-}CCl_3 \longrightarrow Cl : C{-}CCl_3$$

$$Cl_3C{-}C\cdot \frown \cdot C{-}CCl_3 \longrightarrow Cl_3C{-}C : C{-}CCl_3$$

[6.2] TERMINATION: RADICALS REACTING WITH EACH OTHER

Such a reaction of two radicals with each other causes termination of <u>two</u> separate reaction chains; of the potential chain-terminating reactions shown in [6.2], that between two Cl_3C—$CCl_2 \cdot$ radicals is found to be the most common.

The reaction chains involved in the addition of bromine are shorter than those involved in the addition of chlorine, and bromine addition (unlike that of chlorine) is often reversible; the addition of iodine is readily reversible. This ready reversibility has been made use of in the interconversion of isomeric *cis* and *trans* unsaturated compounds:

cis   180° rotation about C — C bond   *trans*

[6.3] INTERCONVERSION OF *cis* AND *trans* UNSATURATED
COMPOUNDS

The radical formed from addition of Br· can either lose Br·, to reform the original *cis* alkene, or undergo 180° rotation about its central C—C bond to yield a species that can then lose Br·, to form the isomeric *trans* alkene. This interconversion is considered further in **10.1.2** (p. 158).

It may well be asked whether there is any evidence that addition of bromine proceeds *via* cyclic bromonium radicals, similar to the cyclic bromonium cations ([5.8], p. 70) observed in electrophilic addition. While there does in general appear to be some preference, under radical conditions, for overall ANTI addition of bromine, this is not sufficiently marked as to suggest the regular participation of such cyclic intermediates.

6.1.2   Addition of HBr

The energetics of hydrogen halide addition to C=C are such that only HBr will add readily *via* a radical pathway. When we come to consider HBr addition to an unsymmetrical alkene such as propene, Me—CH=CH$_2$, we face the problem of orientation, just as we did for electrophilic addition (*cf.* **5.1.3.1**, p. 75). In fact, under radical conditions addition of HBr is found to take place the opposite way round to that observed in electrophilic addition:

1 - bromopropane   2 - bromopropane

[6.4] ORIENTATION: RADICAL VERSUS ELECTROPHILIC ADDITION
OF HBr TO MeCH=CH$_2$

When HBr addition was carried out in preparative terms, however, a mixture of both products was often obtained, the actual composition of the mixture depending on the conditions under which the reaction had been carried out. This caused a great deal of confusion before it was ultimately realised that two different modes of HBr addition could be operating simultaneously.

Reaction under radical conditions requires the provision of a source of radicals to initiate the addition of HBr: organic peroxides, RO—OR, are often used for this purpose, because they will generate RO· under very mild conditions ([4.3], p. 53):

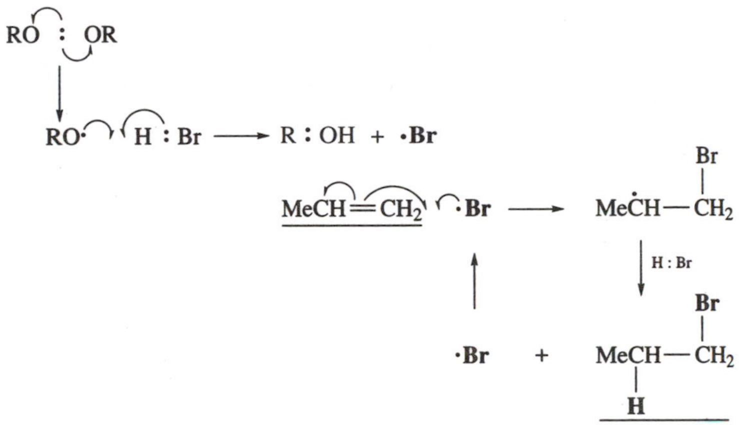

RO : OR

↓

RO· H : Br ⟶ R : OH + ·Br

MeCH═CH₂ ·Br ⟶ MeCH—CH₂
|
Br

↓ H : Br

·Br + MeCH—CH₂
|
H
Br

[6.5] RADICAL ADDITION OF HBr TO MeCH═CH₂

The RO· initiator radical abstracts H· from H—Br to yield Br·, which adds to the double bond to produce a bromoalkyl radical. This new radical can, in turn, abstract H· from a molecule of H—Br to yield the overall addition product, plus a further Br· radical which can repeat the cycle: once again a chain reaction has been set up. The individual reaction chains for HBr addition are, however, relatively short—much shorter than those in halogen addition.

The observed orientation of overall addition will, of course, be controlled by which of the two carbon atoms of the double bond Br· adds to:

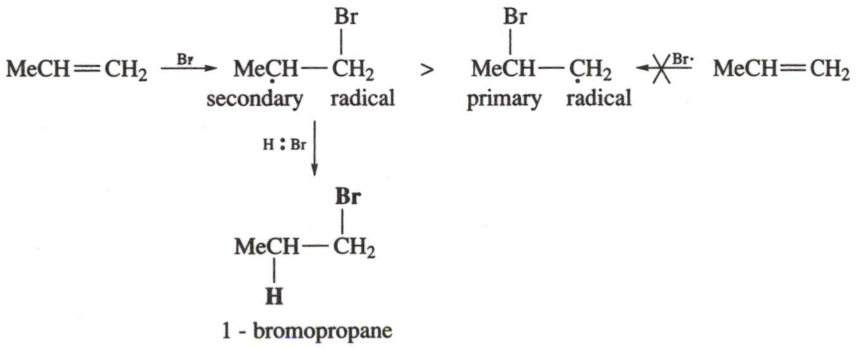

MeCH═CH₂ —Br→ MeCH—CH₂   >   MeCH—CH₂ ←Br·— MeCH═CH₂
| |
Br   Br
secondary radical   primary radical

H : Br ↓

MeCH—CH₂
|
H
Br

1 - bromopropane

[6.6] PREFERRED MODE OF INITIAL ATTACK OF Br· ON MeCH═CH₂

This will be the one that results in the formation of the more stable radical, and as we saw in [4.7] (p. 55) secondary radicals are more stable than primary: hence the observed mode of addition of Br·.

Attack takes place at the double bond carbon atom opposite to the one attacked in the electrophilic reaction, because initial addition was then of $H^{\oplus}$ (*cf.* [5.18], p. 75) to form a carbocation. The determining factor was still the formation of the more stable intermediate: the secondary, rather than the primary, carbocation. Overall addition is then completed, under electrophilic conditions, by addition of $Br^{\ominus}$ to the secondary carbocation intermediate (*cf.* [5.18], p. 75) and, under radical conditions, by abstraction of H· from H—Br by the secondary radical intermediate (*cf.* [6.5], p. 89).

As organic peroxides were often used to initiate addition, the formation of the somewhat unexpected product, 1-bromopropane, was referred to as the **peroxide effect**. The unexpectedness stemmed from the fact that the formation of 1-bromopropane was contrary to Markownikov's rule (**5.1.3.1**, p. 76); it was therefore described as anti-Markownikov addition.

Now that the different possible modes of addition are more clearly understood, it becomes possible to specify the reaction conditions necessary to ensure the preparation of either 1- or 2-bromopropane as required; or indeed to ensure specific Markownikov, or anti-Markownikov, addition of HBr to unsymmetrical alkenes in general. To ensure anti-Markownikov (radical) addition, all that is necessary is to add radical **initiators**, such as peroxides, to the reaction mixture. The chain reaction then set in motion is so much faster than any electrophilic addition, that may be taking place at the same time, as to wholly dictate the composition of the product.

Ensuring Markownikov (electrophilic) addition is a little more difficult because alkenes nearly always contain a small amount of peroxide (arising from their slow reaction with the oxygen in the air (*cf.* **4.2.2**, p. 59)): enough to trigger at least some addition *via* the radical mode. These peroxides are difficult to remove by simple purification, and a more satisfactory technique is to put into the alkene a small quantity of a substance (e.g. certain phenols) which reacts especially readily with radicals, thus removing them from the alkene before HBr addition is attempted. Such species are known as radical **inhibitors** (*cf.* anti-oxidants, **4.2.2**, p. 59).

6.1.3   Vinyl polymerisation

This general reaction, embracing as it does a wide spectrum of unsaturated compounds, has probably received more study than all other radical addition reactions put together; this is because it forms the basis of a large part of the polymer industry.

Just as the carbocation derived from electrophilic addition to an alkene will add to the double bond of another molecule of alkene (**5.1.5**, p. 79), so will a similarly derived radical:

$$Ra \cdot \frown \ \frown CH_2 = CH_2 \longrightarrow Ra : CH_2 - CH_2 \cdot \frown \frown CH_2 = CH_2$$

$$Ra : CH_2 - CH_2 : CH_2 - CH_2 \cdot \xrightarrow{nCH_2 = CH_2} \text{polymer}$$

[6.7] ADDITION OF ALKYL RADICALS TO C=C

The reaction with radicals is usually of greater significance, however, because the reaction chains thus initiated are long, and thus lead to the formation of polymer molecules of high molecular mass. The reaction is generally referred to as **vinyl** polymerisation because the initial unsaturated molecules used (**monomers**) are often of the form, $CH_2 = CH - X$, and $CH_2 = CH -$ is known as a vinyl group. It is usual to consider such polymerisation reactions as involving three phases—**initiation, propagation** and **termination**:

initiation: (a) $RO \overset{\frown}{:} OR \longrightarrow RO \cdot \ \cdot OR$

(b) $RO \cdot \frown \ \frown CH_2 = CH_2 \longrightarrow RO : CH_2 - CH_2 \cdot$

propagation: $RO(CH_2)_2^{\cdot} \xrightarrow{nCH_2 = CH_2} RO(CH_2)_{2n+2}^{\cdot}$

termination: (a) $RO(CH_2)_n \cdot \frown \frown \cdot OR \longrightarrow RO(CH_2)_n : OR$

(b) $RO(CH_2)_n \cdot \frown \frown \cdot_n (CH_2)OR \longrightarrow RO(CH_2)_n : (CH_2)_n OR$

[6.8] INITIATION, PROPAGATION AND TERMINATION IN VINYL POLYMERISATION

Initiation can be effected by any suitable source of radicals, and peroxides are often used for the purpose. As we mentioned previously (**6.1.2**, p. 90) **peroxides** tend to form spontaneously in unsaturated compounds through air oxidation; this can constitute a hazard in stored monomer molecules, which can form tars and gums through auto-polymerisation initiated by radicals formed from these "internal" peroxides. To prevent such premature polymerisation, a small quantity of an inhibitor (**6.1.2**, p. 90) is often inserted into supplies of stored monomers; then, when it is desired to start polymerisation proper, a rather larger amount of peroxide initiator than usual is put in—the excess to "neutralise" any residual inhibitor.

The propagation step is usually extremely rapid, and can involve thousands of molecules of monomer before the reaction chain is terminated. The termination step can involve either of the types of collision shown, but that involving reaction of a growing polymer molecule, $RO(CH_2)_n \cdot$, with an initiator radical, $RO \cdot$, is much the less likely, because all the initiator radicals will have been used up relatively early in the reaction: termination will thus commonly involve collision between two growing polymer molecules.

The length (and hence molecular mass) of the polymer molecules that are produced is not determined solely by the particular monomer that is being used; polymer molecules of widely differing lengths may be produced during the polymerisation of any one particular monomer, depending on the reaction conditions employed.

The physical properties of the solid polymer product are very dependent not only on the length of the molecules in it, but also on the relative proportions of polymer molecules of different lengths that go to make it up. Thus two polymers, in which the <u>average</u> length of the molecule is much the same, will have widely different properties if, in one case, all the molecules are of about the same—average—length, while the other polymer is made up of a mixture of very long and very short molecules. Control of the physical properties of polymers is of such great commerical importance that many ingenious methods have been devised to regulate polymerisation, so as to produce a particular length of molecule, or a particular distribution of molecular lengths, at will.

6.1.3.1   Effect of monomer

The conditions required to induce polymerisation differ considerably from one monomer to another. Thus $CH_2{=}CH_2$ itself requires quite vigorous conditions, including high pressure, to convert it into polythene, while other monomers such as $CH_2{=}CHCl$ ($\rightarrow$polyvinyl chloride, p.v.c), $CH_2{=}CHC_6H_5$ ($\rightarrow$polystyrene), and $CH_2{=}CMeCO_2Me$ ($\rightarrow$perspex) tend to polymerise under rather milder conditions.

Each succesive molecule of $CH_2{=}CHX$ commonly bonds to the growing polymer molecule the same way round, so that the growing polymer molecule adds to the less substituted carbon atom of the next molecule of $CH_2{=}CHX$ monomer:

$$\text{RaCH}_2\!-\!\underset{\overset{|}{X}}{CH^\cdot} \quad CH_2\!\!=\!\!\underset{\overset{|}{X}}{CH} \longrightarrow \text{RaCH}_2\!-\!\underset{\overset{|}{X}}{CH} : CH_2\!-\!\underset{\overset{|}{X}}{CH^\cdot} \xrightarrow{n\,CH_2=\overset{\overset{\textstyle X}{|}}{CH}} \text{polymer}$$

[6.9]   "HEAD-TO-TAIL" POLYMERISATION OF $CH_2{=}CHX$

This is known as "head-to-tail" polymerisation, and most probably results from steric effects, as the CH_2 carbon atom of the monomer molecule will be more accessible to the growing chain than the CHX carbon atom, particularly as X is often bulky. It thus comes as no surpise to find that 1,2-disubstituted alkenes, $XCH{=}CHX$, are exceedingly reluctant to polymerise at all.

In the solid polymer, the substituent X groups will no longer be able to rotate about the carbon–carbon "backbone" of individual polymer molecules: they will be locked in position in the solid. The final relative arrangement of these X substituent groups about the molecular backbone is likely to be largely random, and this will prevent the molecules from lying close together, in an orderly pattern, in the solid polymer: such polymers tend to be non-crystalline, and structurally weak.

Polymerisation catalysts (Ziegler–Natta) have been developed, however, which hold both monomer molecule and growing polymer chain on a molecular template while each addition takes place; this controls the orientation in which the incoming monomer molecule is added. As each CH_2=CHX now adds in an identical, imposed orientation, all the X substituents will be aligned in an orderly arrangement about the molecular "backbone", leading to strong, crystalline polymeric materials.

Another way of influencing the properties of a polymer is by polymerising a mixture of two monomers so that both are incorporated into the individual polymer molecules, either on a 50/50 basis, or in some other proportion—this is known as **copolymerisation**: thus many synthetic rubbers are copolymers of styrene (CH_2=CHC$_6$H$_5$) and buta-1,3-diene (CH_2=CH—CH=CH$_2$). We shall be considering the polymerisation of butadiene itself below (**6.2**, p. 94).

6.1.4   Addition of hydrogen (catalytic)

A very useful reaction in organic synthesis is the direct addition of H_2 to C=C in the presence of finely divided metal catalysts such as nickel, platinum, palladium, and rhodium: **catalytic hydrogenation**. These metals have the common property of being able to adsorb quite large quantities of hydrogen into the surface layers of their metal atoms; they are also capable of adsorbing molecules of alkenes on the metal surface by interaction with the electrons of their double bonds. This is borne out by the observation that both hydrogen and simple alkenes react exothermically (but reversibly) with, for example, nickel.

It seems likely that the H—H bond in the hydrogen molecule is considerably weakened, if not actually broken, in the course of its adsorption into the surface of the metal catalyst; this is the reason for considering hydrogenation under radical addition reactions, even though actual hydrogen atoms as such are not necessarily involved. The double bond in the alkene is also likely to have been broken to some extent, and its electrons made more readily available, when adsorbed on the metal surface.

We can thus envisage a situation in which the alkene molecule, adsorbed in an activated state on the surface of the metal catalyst, is approached by hydrogen atoms (or something rather like them) from layers a little deeper in the metal catalyst as shown in [6.10]:

[6.10] ADDITION OF HYDROGEN TO C=C AT A METAL CATALYST
SURFACE

The reacting species are thus held together on a kind of template while reaction takes place between them; the resulting alkane molecule, lacking available electrons, is not adsorbed by the metal, and is thus released from the catalyst surface, thereby clearing the site for the adsorption of a further molecule of alkene. This desorption of the hydrogenated product from the reaction site is important, as only a very small proportion of the total catalyst surface is found to be "active" enough to effect hydrogenation. This stems from the fact that the spacing of the metal atoms will vary from one face of the metal crystals in the catalyst to another, and only when these spacings approximate to C=C (and/or H—H) bond distances will that face of the metal catalyst crystal constitute an "active" site.

As both hydrogen atoms will have approached the alkene molecule from the same side (from the metal) we might expect the overall addition of hydrogen to be specifically SYN. This is observed sufficiently often as to make hydrogenation stereochemically useful, but is not universal. One reason for this lack of total specificity is that both hydrogen atoms are not usually added to the alkene simultaneously. There may thus be time, after addition of the first H atom, for rapid rotation to take place about the now single carbon–carbon bond of the partially hydrogenated alkene before addition of the second H atom can take place.

Catalytic hydrogenation of C=O, and C=N can also be effected, and that of C≡C is discussed below (**6.4**, p. 100).

6.2   ADDITION TO C=C—C=C

The radical-induced addition reactions of conjugated dienes, such as buta-1,3-diene (CH_2=CH—CH=CH_2), resemble the analogous electrophilic addition reactions in that they proceed *via* a delocalised—and hence stabilised—intermediate, thus making addition somewhat easier than addition to a simple alkene, e.g. CH_2=CH_2.

6.2.1   Addition of halogens

As with electrophilic addition (**5.2.1**, p. 80), the initial radical — in this case
Br·—adds to a terminal carbon atom so that the more stable secondary
(rather than primary) radical intermediate is formed; attack at this position
has the added advantage of leading to the formation of a delocalised
intermediate:

$$
\begin{array}{c}
\overset{\text{Br}}{\underset{|}{\text{CH}_2}}\text{—CH—CH}\text{=}\text{CH}_2 \quad \text{1,2 - addition}
\end{array}
$$

$\dot{\text{C}}\text{H}_2\text{—}\overset{\text{Br}}{\underset{|}{\text{CH}}}\text{—CH}=\text{CH}_2$

$\times$ Br·

$\text{CH}_2=\text{CH—CH}=\text{CH}_2$ $\xrightarrow{\text{Br·}}$

buta - 1,3 - diene

Br : Br

$\overset{\text{Br}}{\underset{|}{\text{CH}_2}}\text{—}\dot{\text{C}}\text{H}\text{—CH}=\text{CH}_2$

$\overset{\text{Br}}{\underset{|}{\text{CH}_2}}\text{—CH}=\text{CH}\text{—}\dot{\text{C}}\text{H}_2$

Br : Br

$\overset{\text{Br}}{\underset{|}{\text{CH}_2}}\text{—CH}=\text{CH}\text{—}\underset{\overset{|}{\text{Br}}}{\text{CH}_2}$ 1,4-addition

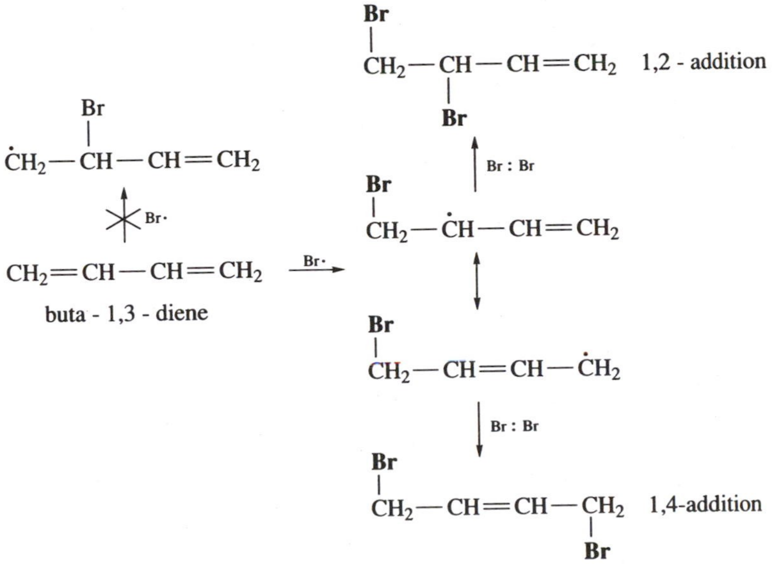

[6.11] 1,2- AND 1,4-RADICAL ADDITION OF Br_2 TO CH_2=CH—CH=CH_2

Attack on this intermediate by Br_2 can then lead to the products of either
1,2- or 1,4-overall addition, or to a mixture of both.

6.2.2   Addition of HBr

As with electrophilic addition (**5.2.2**, p. 81), the question of orientation arises
when HBr adds to an unsymmetrical diene such as MeCH=CH—CH=CH_2.
Two different delocalised intermediates could in theory be formed, as

shown in [6.12]:

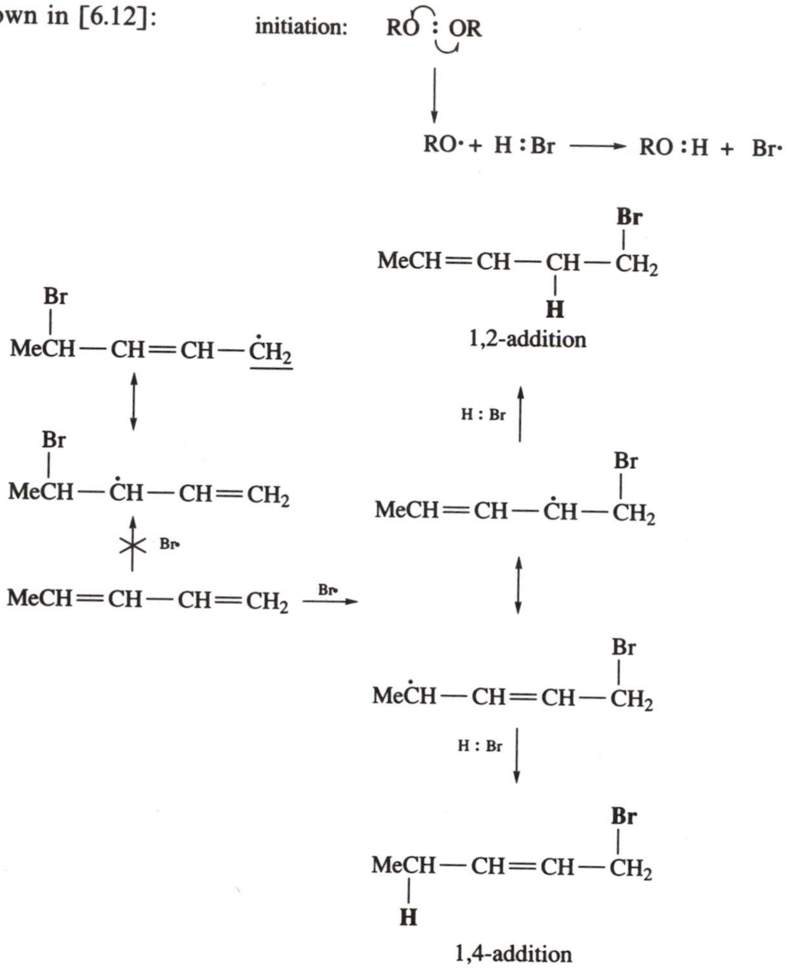

initiation: RO : OR

$$RO· + H:Br \longrightarrow RO:H + Br·$$

$$\underset{\text{1,2-addition}}{MeCH=CH-\underset{\underset{H}{|}}{\overset{\overset{Br}{|}}{CH}}-CH_2}$$

$$MeCH-CH=CH-\dot{C}H_2 \quad \overset{Br}{|}$$

$$MeCH-\dot{C}H-CH=CH_2 \quad \overset{Br}{|}$$

⟶ Br·

$$MeCH=CH-CH=CH_2 \quad \overset{Br·}{\longrightarrow}$$

H : Br ↑

$$MeCH=CH-\dot{C}H-\underset{}{\overset{\overset{Br}{|}}{CH_2}}$$

$$Me\dot{C}H-CH=CH-\underset{}{\overset{\overset{Br}{|}}{CH_2}}$$

H : Br ↓

$$\underset{\underset{H}{|}}{MeCH}-CH=CH-\underset{}{\overset{\overset{Br}{|}}{CH_2}}$$

1,4-addition

[6.12] 1,2- AND 1,4-RADICAL ADDITION OF HBr TO
$$MeCH=CH-CH=CH_2$$

As usual, it is the more stable of the two (the one <u>without</u> any contribution from a primary—underlined—radical structure) that is actually obtained. Attack by HBr can then lead to the product of either 1,2- or 1,4-overall addition, or to a mixture of both. As with a simple unsymmetrical alkene (**6.1.2**, p. 88), overall addition takes place "the other way round" (anti-Markownikov) to electrophilic addition of HBr (**5.2.2**, p. 81).

6.2.3 Polymerisation

The radical-induced polymerisation of conjugated dienes occurs readily, but the resultant polymers offer a new possibility in that they still contain residual double bonds, one from each molecule of diene monomer:

$$Ra \cdot CH_2 = CH - CH = CH_2$$

$$Ra : CH_2 - CH = CH - CH_2 \cdot CH_2 = CH - CH = CH_2$$

$$Ra : CH_2 - CH = CH - CH_2 : CH_2 - CH = CH - CH_2 \cdot$$

$$\downarrow nCH_2 = CH - CH = CH_2$$

polymer

[6.13] RADICAL POLYMERISATION OF CONJUGATED DIENES

Successive molecules of diene monomer are found to add to the growing polymer chain *via* the 1,4-mode, no doubt because it is easier for steric reasons. The residual double bonds in the polymer molecules—one from each monomer unit—can be utilised to modify greatly the physical properties of the bulk polymer.

Such modification may be effected by **"cross-linking"** one polymer molecule to another, using possible reactions of their residual double bonds to form actual chemical bridges between them. Classically, sulphur was used to do this in the vulcanisation of rubber (a naturally occurring polymer) causing its molecules to cross-link through disulphide (S—S) bridges between an initially double bonded carbon atom in one polymer molecule and a similar carbon atom in another.

This cross-linking can occur three-dimensionally in the bulk polymer, and the greater the number of S—S bridges incorporated in this way, the greater the rigidity, and physical strength, of the polymeric material. Thus raw rubber as harvested is a soft and tacky substance, but by vulcanisation it can be transformed, as required, through varying degrees of elasticity to ultimate rigidity, depending on the degree of S—S cross-linking that is incorporated through vulcanisation.

The presence of double bonds in polymer molecules produced from the polymerisation of conjugated dienes, means that the parts of the polymer molecule on each side of such a residual double bond can be either *cis* or *trans*, with respect to each other:

$$\left[\begin{array}{c} CH_2 \quad CH_2 \\ C=C \\ Me \qquad H \end{array} \right]_n \qquad \left[\begin{array}{c} CH_2 \quad H \\ C=C \\ Me \qquad CH_2 \end{array} \right]_n$$

cis (rubber)   *trans* (gutta percha)

[6.14] ALTERNATIVE ORIENTATIONS ABOUT THE RESIDUAL DOUBLE BONDS IN POLYMERISED ISOPRENE

Thus when such a monomer, e.g. isoprene, $CH_2 = C(Me)-CH=CH_2$, is polymerised there can, as shown in [6.14], be either of two different situations about each residual double bond in the polymer chain. It might well be expected that this difference in stereochemistry could have a considerable effect on the physical properties of these two natural polymers; both are derived from the same monomer, isoprene, but rubber has all *cis* junctions, while gutta percha has all *trans*. Such a difference is indeed observed, in that "all *cis*" rubber is—before vulcanisation—a sticky, tacky mess, while "all *trans*" gutta percha is hard and brittle.

6.2.4   Diels–Alder reaction

This is an important reaction of conjugated dienes with suitable compounds containing a carbon–carbon double bond. It is perhaps something of a cheat to consider this reaction here, under the head of radical-induced additions, as it does not actually involve radicals as intermediates; it does, however, proceed under non-polar conditions, and this is the most appropriate place at which to consider it. A typical example is the reaction of buta-1,3-diene with maleic anhydride:

[6.15] DIELS–ALDER REACTION OF BUTA-1,3-DIENE WITH
MALEIC ANHYDRIDE

The Diels–Alder reaction is widespread in its scope, is of considerable synthetic importance (we have, to-date, seen very few reactions in which a new ring system is formed in the product molecule), and is commonly reversible. The simplest possible reaction—that of buta-1,3-diene with ethene—is very slow, and requires vigorous conditions of temperature and pressure.

The Diels–Alder reaction is, in general, facilitated by the presence of electron-donating substituents, e.g. alkyl groups, in the diene, and by the presence of electron-withdrawing groups in the alkene (the **dienophile**), e.g. the $C=O$ groups in maleic anhydride in [6.15]. Reaction is found to take place particularly readily when the two double bonds of the conjugated diene are held in exactly the right orientation—for reaction with the alkene—by being incorporated into a ring structure, as in [6.16]:

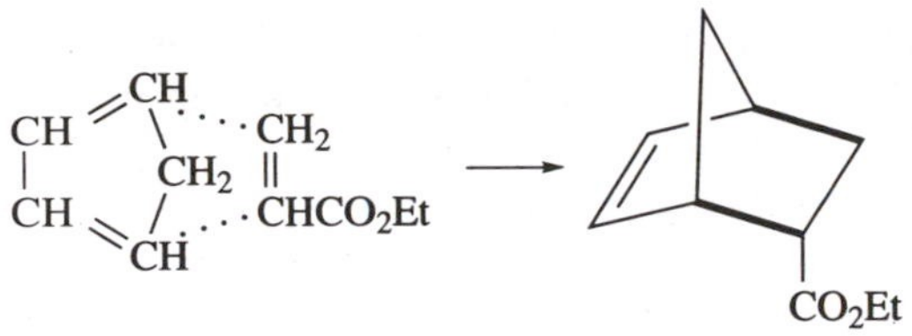

[6.16] DIELS–ALDER REACTION WITH A CYCLIC CONJUGATED DIENE

The Diels–Alder reaction is commonly reversible, under suitable conditions.

6.3 ADDITION TO C=C (aromatic)

Although the major reaction of the halogens, e.g. Cl_2, with benzene was electrophilic <u>substitution</u> (**3.2.2**, p. 36), this did require the presence of Lewis acid catalysts; in their absence, and under conditions that favour the formation of radicals, <u>addition</u> to benzene can also occur:

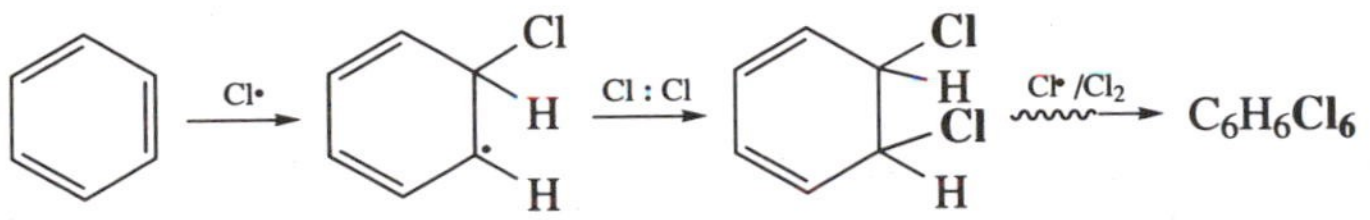

[6.17] RADICAL ADDITION OF Cl_2 TO BENZENE

Thus benzene will react with Cl_2—in light, or in the presence of peroxides, to generate $Cl\cdot$ from Cl_2—to yield the overall addition product, $C_6H_6Cl_6$.

When there is a substituent on the benzene ring that is itself capable of reacting with radicals, e.g. CH_3, then preferential $H\cdot$ abstraction from this by $Cl\cdot$ may well occur (*cf.* **4.2.1.3**, p. 57), rather than the expected addition of the $Cl\cdot$ to a ring carbon atom; such attack will lead to overall <u>substitution</u>:

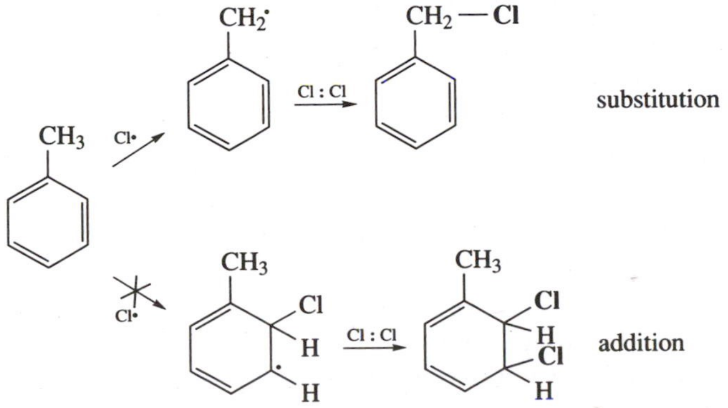

[6.18] PREFERENTIAL RADICAL ATTACK BY $Cl\cdot$ ON CH_3 RATHER THAN ON A BENZENE RING

Preferential abstraction of H· from the CH_3 group by Cl· takes place because addition of Cl_2 to the benzene ring would involve overall loss of aromatic stabilisation, while H· abstraction from the CH_3 group, does not: aromatic character is then retained.

As we have already seen (**4.2.3**, p. 61), overall radical <u>substitution</u> can also be made to take place on a benzene ring, under suitable conditions, e.g. by Ph·.

6.4 ADDITION TO C≡C

Many of the reagents that add to C═C under radical conditions will also add to C≡C; one such reaction, of particular interest and utility, is the addition of hydrogen to alkynes. This can be carried out in the presence of the same catalysts as were used for the hydrogenation of C═C in alkenes, and the product is the alkane resulting from the addition of <u>two</u> molecules of hydrogen. If, however, the addition is carried out in the presence of the **Lindlar** catalyst (finely divided palladium, whose "active sites" have been made less catalytically effective by being "poisoned" through reaction with lead salts) the reaction can be stopped when only one molecule of hydrogen has been added:

$$Me_3CC \equiv CCMe_3 \xrightarrow[\substack{\text{Lindlar} \\ \text{catalyst}}]{H_2} \underset{cis}{\overset{\displaystyle H \qquad H}{\underset{\displaystyle Me_3C \qquad CMe_3}{C=C}}}$$

[6.19] ADDITION OF H_2 TO C≡C IN PRESENCE OF THE
LINDLAR CATALYST

Addition is found to be predominantly SYN, even when—as in [6.19]—this results in the formation of the more crowded *cis* alkene.

6.5 SUMMARY

The electrons of a carbon–carbon double bond, C═C, will also react with the unpaired electron of each of two **radicals** leading to overall **additon**. Many of the species added are the same as those we considered under electrophilic addition to C═C, but whereas the latter required <u>polar</u> conditions, radical addition is promoted by <u>non-polar</u> conditions, light and radical initiators.

The addition of halogens (commonly initiated photochemically) is considered first, and shown to be a very rapid **chain reaction**. Addition of HBr to non-symmetrical alkenes, which requires a radical initiator, is shown to proceed *via* an <u>anti</u>-Markownikov mode, and this is explained in terms of the relative stability of potential intermediates.

Next, vinyl polymerisation is considered in detail, and its industrial importance emphasised; in particular, the effect of different monomers, and of reaction conditions, on the physical properties of the polymer product. Finally the addition of H_2, in the presence of metal catalysts, is discussed.

Radical addition to conjugated systems, C=C—C=C, is now considered, including the possibility of 1,2- and/or 1,4-addition of Br_2, and the orientation of addition of HBr. Polymerisation is then discussed, especially the effect of the residual C=C bonds—one from each monomer unit—on the polymer product. Finally, consideration is given to the Diels–Alder reaction, in which conjugated systems (**dienes**) react with simple C=C compounds containing electron-withdrawing groups (**dienophiles**).

Passing reference is made to <u>addition</u>, in contrast to <u>substitution</u>, in aromatic systems, and to preferential H· abstraction from substituents such as CH_3. Finally, consideration is given to radical addition to C≡C, particularly to the catalytic addition of H_2, which may be terminated at the "half-way" stage by use of the **Lindlar** catalyst.

7

Nucleophilic addition

Addition of an electron-rich nucleophile to a carbon atom requires that such a carbon atom is to some extent electron-deficient; thus a main target for nucleophilic addition is the carbonyl group, C=O, with its $+^{vely}$ polarised carbon atom, $^{\delta+}C{=}O^{\delta-}$ (*cf.* [5.33], p. 84). It is, however, possible to get addition taking place at other unsaturated linkages, e.g. C=C, provided an electron-withdrawing group is attached to one of the carbon atoms of the double bond thus inducing $+^{ve}$ polarity in the other.

7.1 ADDITION TO C=C

7.1.1 Addition to C=C—C≡N (cyanoethylation)

A good example is the way in which the presence of the powerfully electron-withdrawing cyano group, —C≡N, promotes attack by a wide variety of nucleophiles on the non-substituted carbon atom of CH_2=CH—CN (acrylo-nitrile). Thus with ROH, attack on the $+^{vely}$ polarised terminal carbon

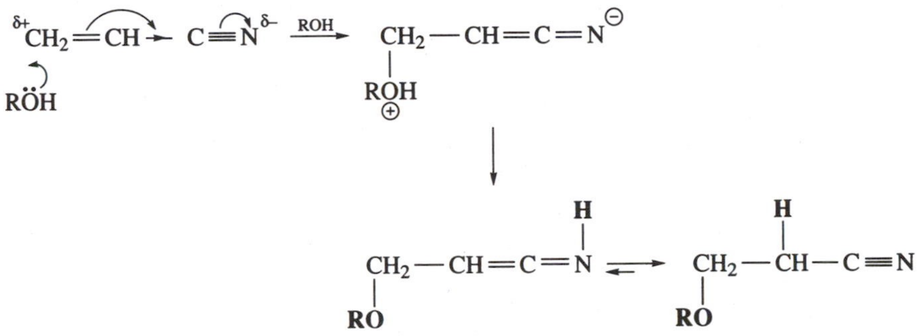

$$^{\delta+}CH_2 = CH \longrightarrow C \equiv N^{\delta-} \xrightarrow{\ ROH\ } CH_2-CH=C=N^{\ominus}$$
$$R\ddot{O}H \qquad\qquad\qquad\qquad\qquad R\underset{\oplus}{O}H$$

$$CH_2-\overset{H}{\underset{RO}{|}}CH=C=N \rightleftharpoons CH_2-\overset{H}{\underset{RO}{|}}CH-C\equiv N$$

[7.1] NUCLEOPHILIC ADDITION TO C=C IN CH_2=CH—CN

atom of CH_2=CH—CN by an electron pair from the O atom of ROH, results in the formation of a bipolar intermediate which can exchange a proton from O to N *via* the solvent. The resultant imino (N—H) compound then isomerises spontaneously to a more stable product in which the H atom is attached to carbon: the product corresponding to net overall addition of ROH to CH_2=CH—.

The reaction is often carried out in the presence of a base, so that the relatively weak nucleophile ROH is converted into the very much stronger one $RO^{\ominus}$. Other potentially useful nucleophiles are H_2O, H_2S, PhOH, RNH_2, etc. As a synthetic tool, the reaction is commonly viewed the other way round: as the attachment to a variety of nucleophiles of the three carbon unit, CH_2=CHCN. The reaction is thus commonly referred to as **cyanoethylation**, and its synthetic utility stems from the possibility of transforming the terminal —CN group into other more useful functional groups, e.g. reduction to CH_2NH_2, or hydrolysis to CO_2H (**7.4**, p. 123).

7.1.2 Addition to C=C—C=O

Although C=O is itself a prime target for nucleophilic attack, it can also act—in the same way as C≡N—as an electron-withdrawing group prompt-ing nucleophilic attack on the other carbon atom of the C=C bond to which it is attached.

7.1.2.1   Addition of HBr

In the addition of HBr to such a system, initial attack is likely to be protonation of the oxygen atom of the C=O group:

$$\text{Me}_2\text{C}=\text{CH}-\underset{\underset{\text{Me}}{|}}{\text{C}}=\text{O} \;\overset{\text{H}^{\oplus}}{\rightleftharpoons}\; \text{Me}_2\text{C}=\text{CH}-\overset{\oplus}{\underset{\underset{\text{Me}}{|}}{\text{C}}}-\overset{\text{H}}{\underset{}{\text{O}}} \longleftrightarrow \overset{\oplus}{\text{Me}_2\text{C}}-\text{CH}=\underset{\underset{\text{Me}}{|}}{\text{C}}-\overset{\text{H}}{\underset{}{\text{O}}}$$

[7.2] ADDITION OF HBr TO Me₂C=CH—C(Me)=O

$$\underset{\text{ketone}}{\text{Me}_2\text{C}-\underset{\underset{\text{Br}}{|}}{\overset{\overset{\text{H}}{|}}{\text{CH}}}-\underset{\underset{\text{Me}}{|}}{\text{C}}=\text{O}} \;\rightleftharpoons\; \underset{\text{enol}}{\text{Me}_2\text{C}-\text{CH}=\underset{\underset{\text{Br}\quad\text{Me}}{|\quad\;|}}{\text{C}}-\overset{\text{H}}{\underset{}{\text{O}}}}$$

The resultant delocalised cation then undergoes attack by Br$^{\ominus}$ at its tertiary cationic centre to yield the enol, Me₂C(Br)CH=C(OH)Me, which isomerises spontaneously to the more stable ketonic structure, Me₂C(Br)CH₂COMe the product corresponding to net overall addition of HBr to Me₂C=CH—. The overall reaction can thus be looked upon as acid-catalysed nucleophilic attack of Br$^{\ominus}$ on the terminal carbon atom of the C=C—C=O system.

Acid-catalysis can also promote the addition of other weak nucleophiles, e.g. ROH, but care has to be taken that the potential nucleophile does not itself undergo significant protonation, as it would then be prevented from acting as a nucleophile because its relevant electron pair would be occupied by H$^{\oplus}$. In many of these simple examples, the alternative of nucleophilic attack on the carbonyl carbon atom is less likely to occur; this is because the products that would thereby be obtained are less stable, and the reactions that would lead to their formation are usually readily reversible, with the equilibrium favouring starting materials rather than products.

7.1.2.2   Addition of RMgBr (Grignard reagents)

Here the position is somewhat less clear-cut in that the product from nucleophilic attack on the carbonyl carbon atom is more stable, and it is not uncommon to get overall addition occurring *via* both 1,4-(conjugate addition, *cf.* **5.2.1**, p. 80) and 1,2-modes:

$$Me_2C=CH-\underset{\underset{Me}{|}}{\overset{\overset{Ph}{|}}{C}}-O^{\ominus} \overset{\oplus}{Mg}Br \xrightarrow[H_2O]{H^{\oplus}} Me_2C=CH-\underset{\underset{Me}{|}}{\overset{\overset{Ph}{|}}{C}}-OH$$

$$\overset{\delta-\ \ \delta+}{Ph\ Mg\ Br}\ \nearrow \qquad\qquad\qquad\qquad\qquad \text{1,2-addition}$$

$$_2C=CH-\underset{\underset{Me}{|}}{C}=O$$

$$\overset{\delta-\ \ \delta+}{Ph\ Mg\ Br}\ \searrow$$

$$Me_2\overset{\overset{Ph}{|}}{C}-CH=\underset{\underset{Me}{|}}{C}-O^{\ominus}\overset{\oplus}{Mg}Br \xrightarrow[H_2O]{H^{\oplus}} Me_2\overset{\overset{Ph}{|}}{C}-CH=\underset{\underset{Me}{|}}{C}-OH \rightleftarrows Me_2\overset{\overset{Ph}{|}}{C}-\underset{\underset{H}{|}}{CH}-\underset{\underset{Me}{|}}{C}=O$$

$$\qquad\qquad\qquad\qquad\qquad\qquad \text{enol} \qquad\qquad\qquad\qquad \text{1,4-addition}$$

[7.3] ADDITION OF PhMgBr TO $Me_2C=CH-C(Me)=O$

The Grignard reagent is itself polarised, and coordination of its metal atom with the oxygen atom of the carbonyl compound makes both the 2- and 4-carbon atoms in $C=C-C=O$ more open to attack by the already powerfully nucleophilic $R^{\delta-}$ (or $Ph^{\delta-}$ in [7.3]) in RMgBr.

Whether overall addition is 1,2- or 1,4- may depend on the relative stability of the alternative products and/or on the steric situation around each of the carbon atoms susceptible to nucleophilic attack. Reactions of C—O with Grignard reagents, and with many other similar organo-metallic reagents (e.g. RLi, R_2CuLi, etc.), are of particular importance because they result in the formation of new carbon–carbon bonds: a highly desirable synthetic activity. The reaction of these reagents with simple carbonyl compounds is discussed in more detail below (**7.2.6.1**, p. 113).

7.2 ADDITION TO C=O

As was said above (p. 102), the major target for nucleophilic addition reactions is the C=O group in carbonyl compounds such as aldehydes, RCHO, and ketones, RCOR.

7.2.1 Effect of structure

Steric effects are found to play a not inconsiderable part in the relative reactivity of carbonyl compounds towards nucleophiles; thus the general reactivity sequence of C=O towards a particular nucleophile is as follows:

$$\underset{H}{\overset{H}{>}}C\!\!=\!\!O \quad > \quad \underset{H}{\overset{R}{>}}C\!\!=\!\!O \quad > \quad \underset{R}{\overset{R}{>}}C\!\!=\!\!O$$

[7.4] STERIC EFFECT OF SUBSTITUENTS ON EASE OF REACTION OF C=O

This results from the decreasing ease of access to the carbonyl carbon atom as the <u>size</u> of the substituents (and also of the attacking nucleophile) increases; but, more particularly, from the resultant increase in the degree of crowding about this carbon atom as its state changes from planar (in the original C=O) to tetrahedral (in the forming intermediate):

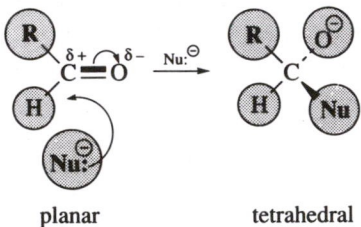

planar   tetrahedral

[7.5] CHANGE AT CARBONYL CARBON ATOM ON NUCLEOPHILIC ATTACK

Electronic effects will also play a part in that electron-donation by alkyl substituents—though not large—will serve to decrease the $+^{ve}$ polarisation of a carbonyl carbon atom to which they are attached, thereby making nucleophilic attack on it more difficult:

$$\underset{H}{\overset{H}{>}}\overset{\delta+}{C}\!\!=\!\!\overset{\delta-}{O} \quad > \quad \underset{H}{\overset{Me}{>}}\overset{\delta+}{C}\!\!=\!\!\overset{\delta-}{O} \quad > \quad \underset{Me}{\overset{Me}{>}}\overset{\delta+}{C}\!\!=\!\!\overset{\delta-}{O}$$

[7.6] ELECTRONIC EFFECT OF SUBSTITUENTS ON EASE OF REACTION OF C=O

It is also found that C_6H_5CHO is markedly less reactive than, for example, CH_3CHO, under comparable conditions. A phenyl group, C_6H_5, is certainly larger than a methyl group, CH_3, and might thus be expected to offer correspondingly greater hindrance to the approach of an attacking nucleophile; C_6H_5 is, however, flat and could to some extent counteract the effect of its bulk by twisting out of the way of the approaching nucleophile.

The major reason for the observed difference in reactivity is that the benzene ring stabilises the C=O group through electronic interaction (delocalisation):

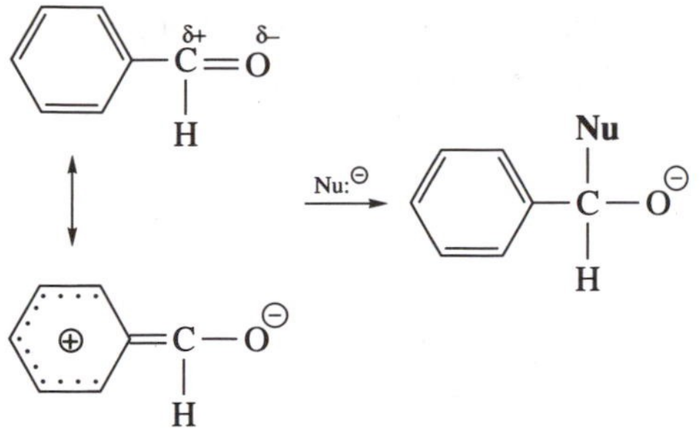

[7.7] ELECTRONIC STABILISATION OF C=O BY C_6H_5

This stabilisation is lost, progressively, as the tetrahedral intermediate is being formed during nucleophilic attack, thus making such attack occur less readily.

Another important point about nucleophilic addition to C=O is that such reactions (apart from those with extremely powerful nucleophiles such as $H^\ominus$) are usually reversible. Thus apart from the ease, or otherwise, with which a particular addition will take place, we also need to know how far over—in favour of the product—the position of equilibrium of the reaction lies. In many cases, changing the structure of either carbonyl compound or nucleophile is found to influence both rate of reaction, and position of equilibrium, in the same direction: the faster a reaction is, the further over in favour of the product its position of equilibrium is also. These points are well illustrated in the addition of H_2O to C=O.

7.2.2   Addition of H_2O

The addition of H_2O to $H_2C=O$, MeHC=O and $Me_2C=O$ demonstrates clearly the effect of progressive substitution at the carbonyl carbon atom on the position of equilibrium for reaction with the same nucleophile:

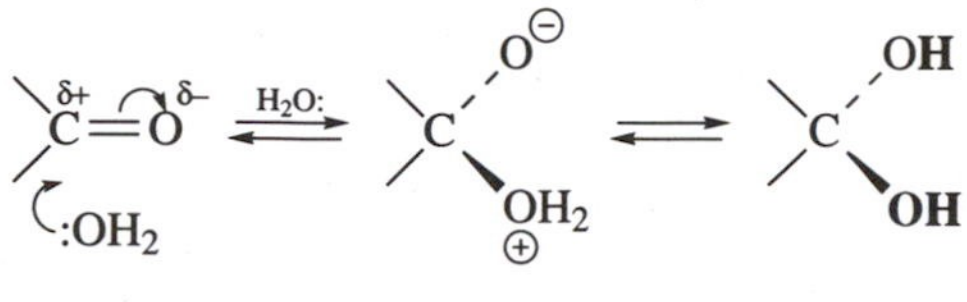

	% hydration at equilibrium
$H_2C=O$	≈ 100
MeHC=O	58
$Me_2C=O$	≈ 0

[7.8] EFFECT OF SUBSTITUENTS ON DEGREE OF HYDRATION OF C=O

Reaction with $H_2C=O$ results in essentially complete hydration, but the easy reversibility of the reaction is reflected in the fact that the unhydrated aldehyde is recovered in 100% yield on warming an aqueous solution of the hydrate. It may be shown that reaction does indeed take place with $Me_2C=O$—even though the equilibrium concentration of hydrate is essentially zero—by carrying out the reaction in water whose oxygen atom has been "labelled" with the heavier oxygen isotope, ^{18}O:

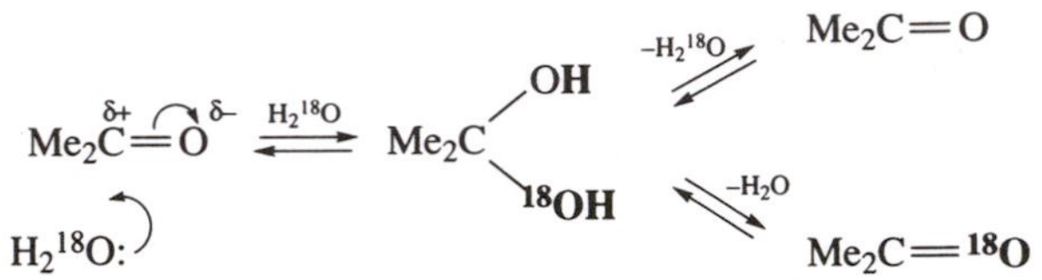

[7.9] REACTION OF $Me_2C{=}O$ WITH $H_2{}^{18}O$

Addition of $H_2{}^{18}O$ would lead to a hydrate containing both OH and ^{18}OH groups, and reversal of this hydration—in the course of the reaction's dynamic equilibrium—could involve loss either of OH or of ^{18}OH. Assuming that loss of OH and ^{18}OH occur with equal readiness (more or less), the result, over a period of time, should be a growing proportion of ^{18}O isotope in the $Me_2C{=}O$: which is exactly what is indeed observed. This exchange is found to take place only very slowly in $H_2{}^{18}O$ itself at pH 7, but very rapidly in the presence of a trace of acid as catalyst.

With a few rather special carbonyl compounds it has actually proved possible to isolate stable hydrates, as opposed to only detecting them spectroscopically (e.g. as with the hydrate of $MeHC{=}O$). The possibility of isolation requires the presence, in the original carbonyl compound, of structural features which can specifically stabilise the hydrate:

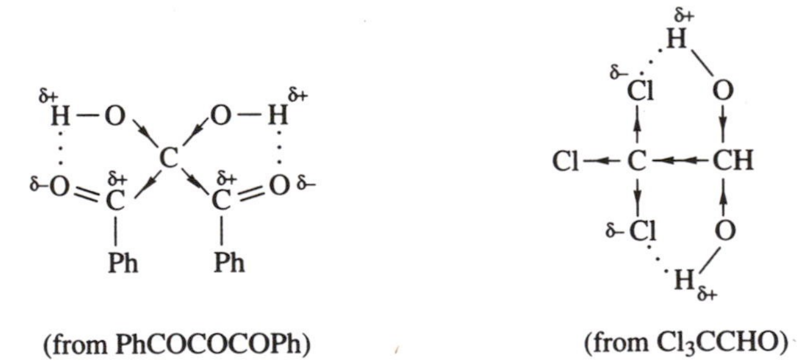

(from PhCOCOCOPh)   (from Cl_3CCHO)

[7.10] STABILISATION OF CARBONYL HYDRATES

In the first of the two examples in [7.10]—the product obtained on hydration of PhCOCOCOPh—stabilisation is promoted by the electron-withdrawing effect of the two·adjacent $C{=}Os$ on the OH groups now attached to the central carbonyl carbon atom. Even more important in promoting stabilisation, however, is the **hydrogen bonding** that can occur between the H atoms of the two OH groups and the electronegative O atom of the two adjacent carbonyl groups.

The second example in [7.10]—the product obtained on hydration of Cl_3CCHO—is very similar: the electron-withdrawing effect of the three Cl atoms, and hydrogen bonding between the H atoms of the OH groups and these electonegative Cl atoms, both acting to stabilise the hydrate. The original aldehyde (Cl_3CCHO) is an oily liquid which, on pouring into water is converted into a crystalline hydrate (chloral hydrate).

7.2.3   Addition of ROH, RSH

Hardly surprisingly alcohols, e.g. EtOH, react with C=O in the same way as H_2O, but being somewhat weaker nucleophiles the reaction tends to proceed somewhat less readily: the product is a <u>hemi-acetal</u>, e.g. MeCH(OH)OEt:

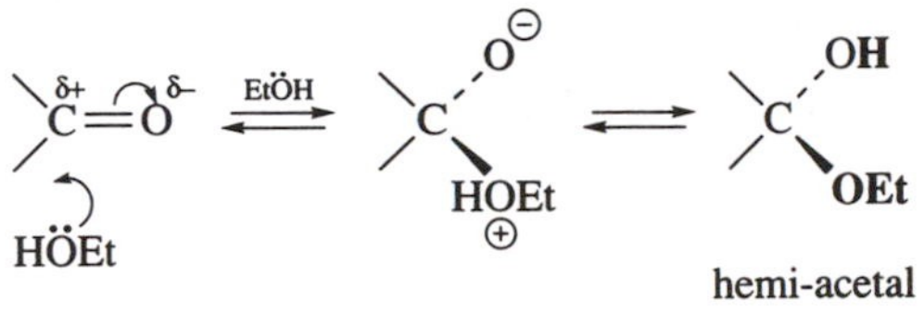

[7.11] FORMATION OF HEMI-ACETALS FROM EtOH AND C=O

As with hydrates, electron-withdrawing substituents on the carbonyl carbon atom will stabilise a hemi-acetal, and may—in special cases—allow its isolation.

If an acid-catalyst is present the reaction can proceed further to yield an <u>acetal</u>, e.g. $MeCH(OEt)_2$:

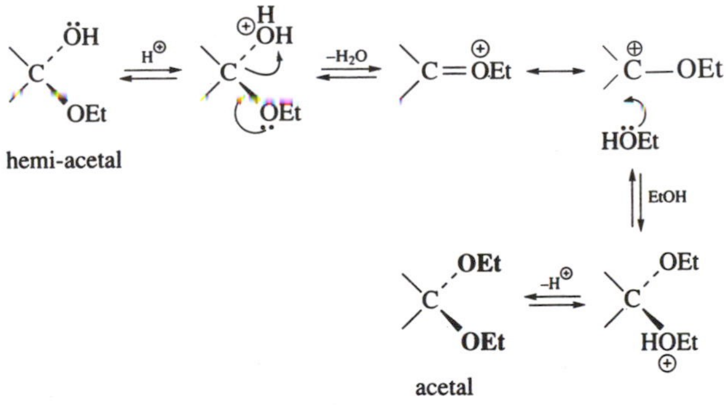

[7.12] FORMATION OF ACETALS FROM HEMI-ACETALS

The overall reaction from carbonyl compound to acetal is reversible, and the position of equilibrium will be influenced by the nature of both the alcohol and the carbonyl compound; thus the reaction proceeds poorly, if at all, with ketones, RCOR.

The equation for acetal formation can be written as,

[7.13] EQUATION FOR ACETAL FORMATION

from which it may be seen that the position of equilibrium may be driven over to the right-hand side—in favour of the acetal product—by using a large excess of the alcohol, e.g. EtOH (the relevant alcohol is indeed commonly used as the solvent for the reaction), and also by removing the H_2O as it is formed—which can be achieved in a number of convenient ways.

Acetal formation may be used for the **protection** of suitable carbonyl compounds, as reactions may then be carried out on the X part of $XCH(OEt)_2$, e.g. oxidation or reduction, which otherwise would also attack the CHO group in unmodified XCHO; thus acetals are stable to bases, which react particularly readily with their parent aldehydes (*cf.* **7.2.6.3**, p. 115). Another essential requirement of a protecting group is, of course, that it should be easily removeable once the danger is over! This presents no problems as acetals are readily hydrolysed back to the original carbonyl compound by dilute aqueous acid:

$$\diagdown\!\!\!\diagup C(OEt)_2 \; + \; excess \; H_2O \;\overset{H^\oplus}{\rightleftharpoons}\; \diagdown\!\!\!\diagup C{=}O \; + \; 2EtOH$$

[7.14] REVERSAL OF ACETAL FORMATION

This provides the necessary acid-catalysis, and also an excess of H_2O which will drive the position of equilibrium over to the right hand side—in favour of the original carbonyl compound (*cf.* **9.3.1**, p. 147).

Thiols, RSH, are rather more powerful nucleophiles than alcohols, ROH, because sulphur is a less electronegative atom than oxygen; it thereby exerts a slightly less tight grip on its outer electrons, which are thus more readily available to bond with another atom or group. This is reflected in the fact that thiols, e.g. EtSH, will normally form <u>thioacetals</u> with ketones as well as with aldehydes:

$$\diagdown\!\!\!\diagup C{=}O \; + \; 2EtSH \;\rightleftharpoons\; \diagdown\!\!\!\diagup C(SEt)_2 \; + \; H_2O$$

[7.15] FORMATION OF THIOACETALS

Thioacetals, in contrast to acetals, are relatively stable towards hydrolysis by dilute acid, but can be reconverted to the parent carbonyl compound on treatment with aqueous $HgCl_2$ solution:

$$\diagdown\!\!\!\diagup C(SEt)_2 \; + \; H_2O/HgCl_2 \;\longrightarrow\; \diagdown\!\!\!\diagup C{=}O \; + \; Hg(SEt)_2 \; + \; 2HCl$$

[7.16] HYDROLYSIS OF THIOACETALS

It seems likely that $Hg^{2\oplus}$ removes EtSH irreversibly, as it is produced, thereby driving the position of equilibrium over to the right-hand side—in favour of the parent carbonyl compound. With both acetal and thioacetal formation available, we have access to a broad spectrum of the potential protection of carbonyl compounds.

Another useful synthetic application of thioacetals is that they may be desulphurised by treatment with excess of an active nickel catalyst (Raney nickel, *cf.* **6.1.4**, p. 93):

$$\text{\textbackslash}C{=}O \xrightleftharpoons{\text{EtSH}} \text{\textbackslash}C(\text{SEt})_2 \xrightarrow[\text{[H}_2\text{]}]{\text{Ni}} \text{\textbackslash}CH_2$$

[7.17] C=O → CH$_2$ VIA DESULPHURISATION OF THIOACETALS

The nickel catalyst contains adsorbed hydrogen (produced during its preparation), which it can transfer to the carbon atom of the thioacetal, whose sulphur atoms have been irreversibly bonded to the nickel surface. The usefulness of the reaction lies in the possibility of direct reduction of C=O to CH$_2$ under relatively mild conditions, which is otherwise difficult to do (*cf.* [3.18], p. 40).

7.2.4   Addition of H$_2$ via metal hydride ions

Hydride ion, H$^\ominus$, can be considered as the anion that would be obtained if H$_2$ were to act as an acid! H$_2$ is, of course, an extremely weak acid, and its anion, H$^\ominus$ will, therefore, be an extremely strong base. It will also be a correspondingly powerful nucleophile; so much so that its reaction with C=O is found to be essentially irreversible. H$^\ominus$ is not normally used as such (as, for example, sodium hydride, Na$^\oplus$H$^\ominus$), but in the form of complex metal hydrides, e.g. lithium aluminium hydride, Li$^\oplus$AlH$_4^\ominus$, sodium borohydride, Na$^\oplus$BH$_4^\ominus$, and many others, which can act as donors of H$^\ominus$:

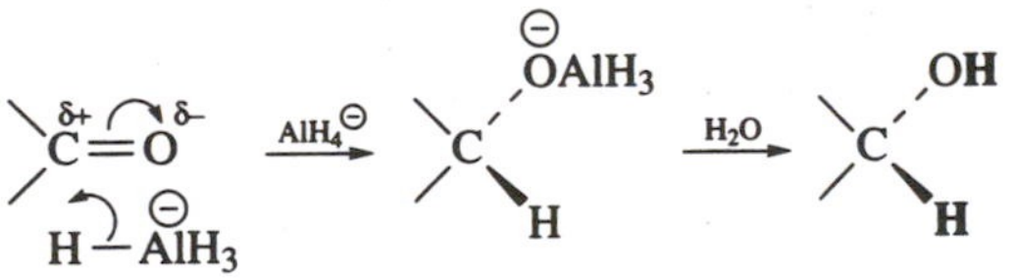

[7.18] REDUCTION OF C=O WITH LiAlH$_4$

A molecule of LiAlH$_4$ has <u>four</u> available H atoms and is thus capable of reducing the C=O groups of <u>four</u> molecules of carbonyl compound, not merely the one as shown here.

The various metal hydrides differ considerably in their reactivity. Thus reductions with LiAlH$_4$ cannot be carried out in hydroxylic solvents as the hydride is decomposed by ROH with considerable violence; this stems from LiAlH$_4$ abstracting hydrogen from ROH—as H$^\oplus$—to form H$_2$. By contrast, NaBH$_4$ is not decomposed in the same way, and may thus be used to effect reductions in hydroxylic solvents.

These hydrides exhibit a considerable degree of specificity as reducing agents, in that while they will reduce C=O in almost any situation (e.g. LiAlH$_4$ will reduce the C=O group in esters, RCO$_2$Et, to RCH$_2$OH) they have no effect on C=C (except in some cases of C=C—C=O), or on C≡C. Carbonyl groups may also be reduced with H_2 in the presence of suitable catalysts (*cf.* **6.1.4**, p. 93).

7.2.5   Addition of HCN

This was the first nucleophilic addition to C=O to be studied in sufficient detail to allow the establishment of a general mechanism for this type of reaction (it was indeed the first organic reaction mechanism to be studied at all!):

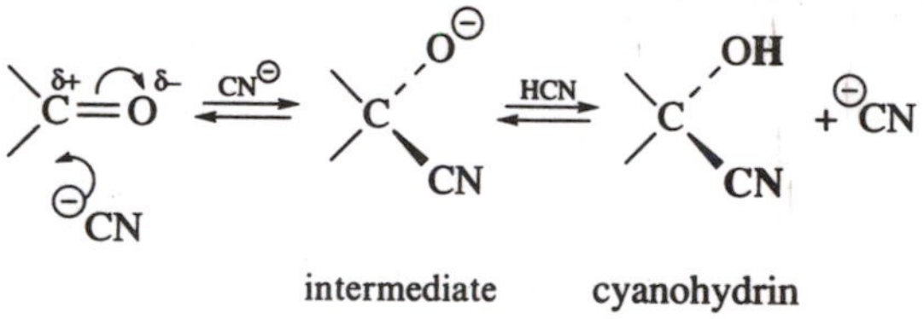

intermediate   cyanohydrin

[7.19] ADDITION OF HCN TO C=O

The product is a **cyanohydrin**, and the reaction is found to be reversible, the position of equilibrium in a particular case being sensitive to the pattern of substitution about the carbonyl carbon atom. Thus the reaction is found to be preparatively useful with most aldehydes, and with simple ketones (those with small R groups).

Reaction with HCN itself is found to be poor; this is because HCN is so weak an acid that it provides only a very low concentration of the required nucleophile, $^{\ominus}$CN. The reaction is greatly speeded up by the addition of a base, so as to produce more $^{\ominus}$CN from HCN, but KCN itself (i.e. $^{\ominus}$CN alone) does not work as there is then no source of proton to convert the intermediate into the product cyanohydrin, thereby pulling the equilibrium over to the right-hand side. The best preparative conditions appear to be to use an excess of KCN plus one mole of strong acid, thus having the best of both worlds.

The importance of the reaction lies not only in the formation of a carbon–carbon bond, but also in the fact that the added CN group is transformable into other, more useful, functional groups:

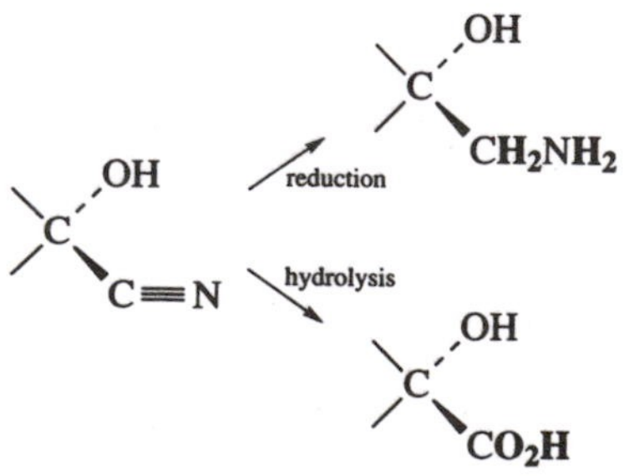

[7.20] TRANSFORMATION OF CYANOHYDRIN —CN INTO OTHER
FUNCTIONAL GROUPS

7.2.6 Addition of carbon nucleophiles

The major importance of adding species containing a nucleophilic carbon atom to C=O is that a new carbon–carbon bond is thereby formed; the enormous variety of such nucleophilic species makes this a general synthetic reaction of very great flexibility, and hence utility. The simplest of such species is the negatively polarised carbon atom in organo-metallic compounds, e.g. Grignard reagents.

7.2.6.1 RMgBr etc.

We have already mentioned this reaction, in passing, when considering addition to C=C—C=O (**7.1.2.2**, p. 104). Addition of organo-metallic compounds such as Grignard reagents to C=O, is a useful method for

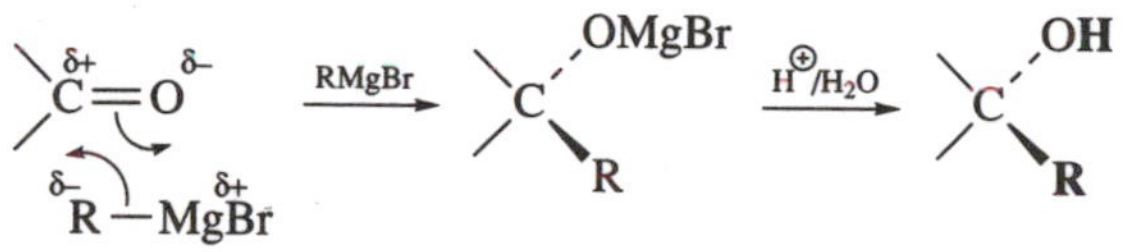

[7.21] ADDITION OF RMgBr TO C=O

forming C—C bonds, because of the wide variety of alkyl, and aryl, groups (in RMgBr) that may be employed, and also of the range of carbonyl compounds with which they can be made to react. In cases where reaction is slow, it may often be speeded up through the use of a mild Lewis acid catalyst, e.g. $MgBr_2$, which complexes with the oxygen atom of the C=O group thereby increasing the $+^{ve}$ polarity of its carbonyl carbon atom:

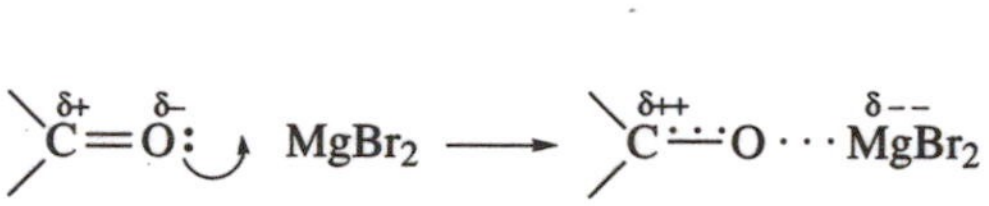

[7.22] USE OF LEWIS ACID CATALYST TO INCREASE REACTIVITY
OF C=O

The reaction of C=O with Grignard reagents is essentially irreversible, and one of the few limitations on it is that it cannot be carried out in hydroxylic solvents as these decompose Grignard reagents on contact with them. The simple addition product from C=O is an alcohol:

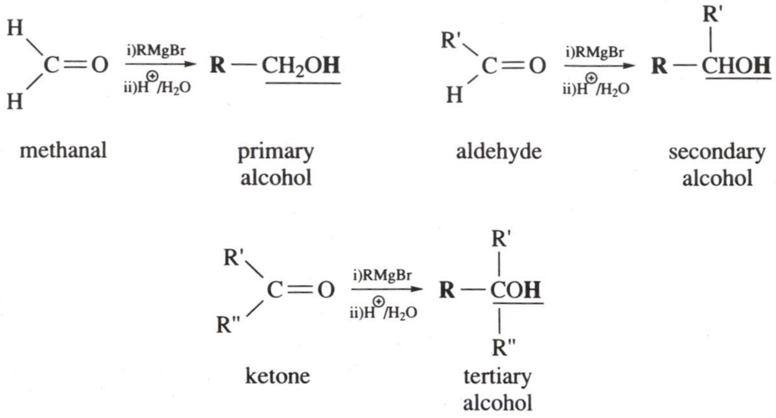

methanal → primary alcohol; aldehyde → secondary alcohol

ketone → tertiary alcohol

[7.23] ALCOHOLS FROM RMgBr/C=O

The unique (unsubstituted) methanal, CH_2O, yields a primary alcohol— RCH_2OH; aldehydes, R′CHO, yield secondary alcohols, $\overline{RR'CHOH}$; while ketones, R′R″CO, yield tertiary alcohols, $\overline{RR'R''COH}$. It should, however, be emphasised that this is not just a general method for synthesising alcohols; the OH group that these contain may readily be replaced by other functional groups, e.g. Br, which can then be involved in further synthetic activity. Alternatively, the alcohols can be dehydrated (*cf.* **9.1.1**, p. 142) to yield C=C, which can then be the subject of further synthetic activity through addition reactions (*cf.* **5.1**, p. 67).

For a number of carbon–carbon bond-forming purposes, Grignard reagents have been superseded, as nucleophilic adducts to C=O, by a wide variety of other organo-metallic reagents, e.g. RLi, PhLi, R_2CuLi, which work better than RMgBr in particular situations.

7.2.6.2 Acetylide anions

Another useful synthetic procedure, though much less flexible than the use of Grignard reagents, is the addition to C=O of the anions derived from ethyne (acetylene), HC≡CH, and its mono-substituted derivatives, RC≡CH. We have already noted that acetylene is markedly acidic (**5.3**, p. 82), and strong bases, e.g. $^{\ominus}NH_2$, will convert it (and RC≡CH) into their anions:

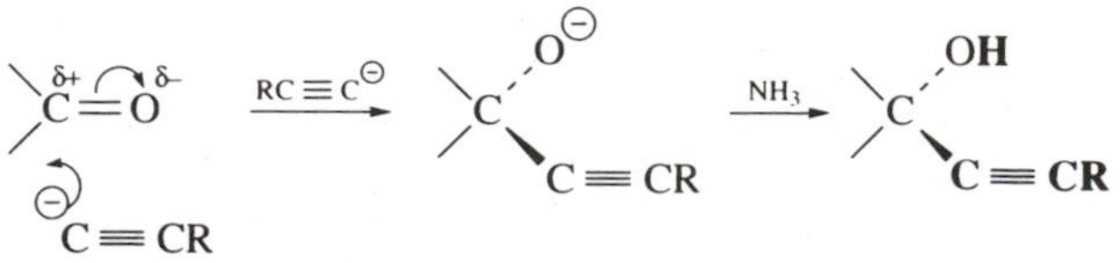

[Reaction scheme continued on p. 115]

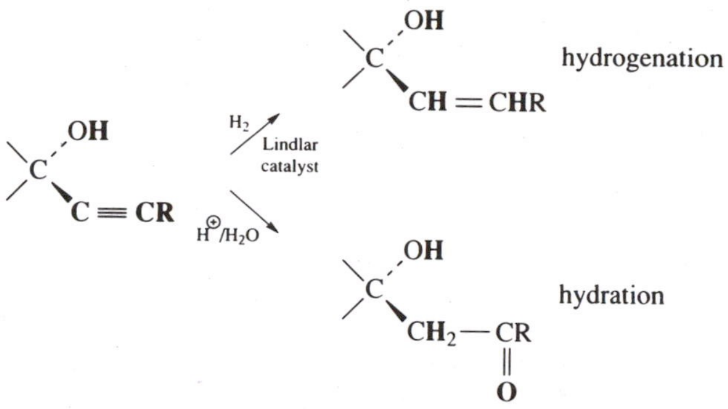

$$\text{hydrogenation} \qquad \overset{OH}{\underset{CH=CHR}{C}}$$

$$\overset{OH}{\underset{C\equiv CR}{C}} \quad \xrightarrow[\substack{H_2 \\ \text{Lindlar} \\ \text{catalyst}}]{} $$

$$\overset{OH}{\underset{CH_2-CR}{\underset{\underset{O}{\|}}{C}}} \qquad \text{hydration}$$

[7.24] ADDITION OF RC≡C⊖ TO C=O AND TRANSFORMATIONS
OF THE —C≡CR ADDUCT

This addition can conveniently be carried out in liquid ammonia, in which ⊖NH₂ can readily be generated—as Na⊕NH₂⊖—through reaction of the NH₃ with metallic sodium. The triple bond in the addition product can then be hydrogenated to C=C by use of the Lindlar catalyst (**6.4**, p. 100), hydrated to form a ketone (*cf.* **5.3**, p. 83), or modified in other ways, leading to new functional groups which can then be involved in further synthetic activity.

7.2.6.3 Carbanions from carbonyl compounds

When carbonyl compounds are treated with aqueous base, there is little doubt that ⊖OH can act as a nucleophile, and add to the carbon atom of the C=O group:

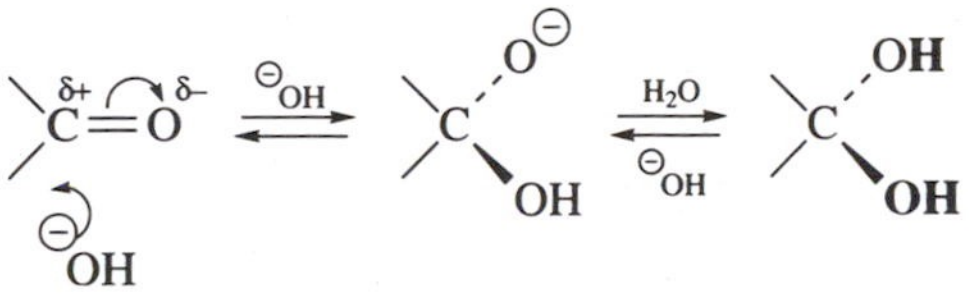

[7.25] ADDITION OF ⊖OH TO C=O

Overall addition—of H₂O—could then be completed by abstraction of a proton from a molecule of the solvent (H₂O), but such overall hydration is extremely readily reversible (*cf.* **7.2.1**, p. 107) and does not lead to any useful synthetic result.

Where, however, the C=O group is attached to a carbon atom that has at least one H substituent (e.g. —CHCHO, —CHCOR, —CHCO₂Et), then electron-withdrawal by the C=O group results in such H atoms being acidic:

$$HO^{\ominus} \curvearrowright \overset{H}{\underset{H}{\underset{|}{C}}}-C=O \underset{H_2O}{\overset{\ominus OH}{\rightleftharpoons}} \overset{}{\underset{H}{C}}-\overset{\ominus}{C}=O \longleftrightarrow \overset{}{C}=\overset{}{\underset{H}{C}}-\overset{\ominus}{O}$$

delocalised carbanion

[7.26] ACIDITY OF H ON A CARBON ATOM ATTACHED TO C=O

An alternative reaction open to aqueous base would thus be to remove such an acidic H atom, thereby setting up an equilibrium with the carbanion derived from the original carbonyl compound. Such a carbanion will be stabilised through delocalisation, and could act as a nucleophile towards the C=O group in molecules of the same, or of a different, carbonyl compound.

7.2.6.3.1   Carbanions from aldehydes

The carbanion derived from the action of an aqueous base on a simple aldehyde such as CH_3CHO (ethanal) can thus add to the C=O group of a molecule of CH_3CHO that has not lost a proton, thereby forming a hydroxy aldehyde as the product of addition:

$$Me-\overset{H}{\underset{\underset{\ominus CH_2CHO}{}}{\overset{|}{C}}}\overset{\delta+ \curvearrowright \delta-}{=}O \underset{\ominus CH_2CHO}{\rightleftharpoons} Me-\overset{H}{\underset{CH_2CHO}{\overset{|}{\underset{|}{C}}}}-O^{\ominus} \underset{H_2O}{\rightleftharpoons} Me-\overset{H}{\underset{CH_2CHO}{\overset{|}{\underset{|}{C}}}}-OH$$

[7.27] ADDITION OF $^{\ominus}CH_2CHO$ TO CH_3CHO: ALDOL REACTION

The equilibrium that is set up with aldehydes such as ethanal is usually found to lie well over on the right-hand side, in favour of the addition product. This addition to C=O is known as the **aldol** reaction.

The hydroxy aldehyde product still contains an acidic H atom (on the carbon atom adjacent to its C=O group), and is thus itself capable of forming a new carbanion on treatment with aqueous base:

$$HO^{\ominus}\curvearrowright \overset{H}{\underset{OH}{\overset{|}{MeCH}}}-\overset{|}{CHCHO} \underset{(1)}{\overset{\ominus OH}{\rightleftharpoons}} \overset{}{\underset{OH}{\overset{}{MeCH}}}-\overset{\ominus}{CHCHO}$$

carbanion

$$\overset{\ominus}{OH}\Big\updownarrow (2)$$

$$\underset{\overset{|}{\overset{\ominus}{OH}}}{MeCH}-\overset{\ominus}{CHCHO} \overset{\ominus OH}{\rightleftharpoons} MeCH=CHCHO$$

carbanion

$$\overset{H}{\underset{OH}{\overset{|}{MeC}}}\overset{\delta+\curvearrowright\delta-}{=}O \overset{MeCHO}{\rightleftharpoons} \overset{H}{\underset{OH}{\overset{|}{MeC}}}-O^{\ominus} \Big\updownarrow H_2O$$

$$\overset{H}{\underset{}{\overset{|}{MeC}}}-OH$$
$$\underset{\underset{OH}{|}}{MeCH}-CHCHO$$

polymer

[7.28] FURTHER REACTIONS OF PRODUCT OF INITIAL ALDOL ADDITION

This new carbanion can now add to the C=O group of a further molecule of ethanal, and the product of this addition can, in turn, form a carbanion, and so the cycle can go on repeating itself. The end result from the action of a strong base on such aldehydes is thus sticky, low molecular-mass polymers. The reaction can, however, be stopped—after the first simple addition—by using weaker bases such as K_2CO_3.

An alternative reaction—of the carbanion derived from the initial hydroxy aldehyde—is loss of $^{\ominus}OH$ (i.e. overall dehydration, *cf.* [8.13], p. 133), resulting in the formation of an unsaturated aldehyde in which the introduced C=C is conjugated($\alpha\beta$-) with the original C=O: this dehydration reaction does also take place. Dehydration reactions induced by a base are unusual, loss of H_2O commonly proceeding only under acid-catalysed conditions (*cf.* **9.1.1**, p. 142).

"Mixed" aldol reactions, involving two different—though similar—aldehydes, are not commonly of any preparative value. This stems from the fact that interaction of the two aldehydes with the two carbanions derived from them could (and generally does) lead to a mixture of four different products. Some "mixed" aldol reactions can, however, be made to work: as, for example, when one of the aldehydes is unable to form a carbanion, e.g. PhCHO, and can thus only be the recipient of the carbanion derived from the other aldehyde:

[7.29] ALDOL REACTION OF $^{\ominus}CH_2CHO$ WITH PhCHO

No aldol reaction can take place with aldehydes of the form HCHO, PhCHO or R_3CCHO alone, because none of them is capable of forming a carbanion; reaction of any of these aldehydes with aqueous base could thus result merely in addition of $^{\ominus}OH$ to their C=O groups. If, however, the base is a strong one, and if it is present in high concentration, then further reaction is indeed found to take place with such aldehydes, e.g. with PhCHO:

[7.30] CANNIZZARO REACTION OF PhCHO WITH KOH

The product of initial $^{\ominus}OH$ addition is found to transfer H with an electron pair—i.e. as the very powerful nucleophile hydride ion ($H^{\ominus}$)—to the carbonyl carbon atom of a second molecule of PhCHO, thus forming the pair of molecules: $PhCO_2H$, and the anion $PhCH_2O^{\ominus}$. This pair will then exchange a proton, *via* the aqueous solvent, to form the more stable pair of molecules: $PhCO_2^{\ominus}$, and $PhCH_2OH$. This is known as the **Cannizzaro** reaction, and is overall a disproportionation: one molecule of PhCHO has been oxidised to $PhCO_2^{\ominus}$, while the other has been reduced to $PhCH_2OH$.

7.2.6.3.2   Carbanions from ketones

The carbanion derived from a simple ketone such as propanone, CH_3COCH_3, will also add to the C=O group of another molecule of the same ketone; but with ketones the position of equilibrium is commonly found to be

[7.31] ADDITION OF $^{\ominus}CH_2COCH_3$ TO CH_3COCH_3

well over on the left-hand side of the equilibrium—in favour of the starting materials. This reflects the greater difficulty of attack on "keto" C=O, compared with attack on "aldehydo" C=O, in the parallel reaction with, for example, CH_3CHO (*cf.* **7.2.1**, p. 105).

There is however, an ingenious way in which the yield of the reaction with propanone can be improved: preventing the addition product, once formed, from having further contact with the base catalyst, thereby ruling out any possibility of it reverting, *via* the "back" reaction, to starting materials.

This can be done by boiling the propanone under a reflux condenser, and arranging for the ketone vapour—after condensation—to trickle back into the distillation flask over a solid base catalyst, $Ba(OH)_2$. Some conversion of propanone into addition product thereby takes place, but whereas unreacted propanone is "recycled" over the catalyst for as long as refluxing is continued, the much less volatile addition product remains in the distillation flask, away from further contact with the catalyst, and so does not revert to starting material. The proportion of addition product in the flask thus increases progressively, and refluxing is continued until an acceptable yield of product has been obtained.

7.2.6.3.3   Carbanions from esters

Carbanions may also be obtained from simple esters such as ethyl ethanoate, CH_3CO_2Et ($EtO^{\ominus}$, from $EtO^{\ominus}Na^{\oplus}$, is commonly used as the necessary base catalyst), and these can then add to the C=O group of a second molecule of the ester:

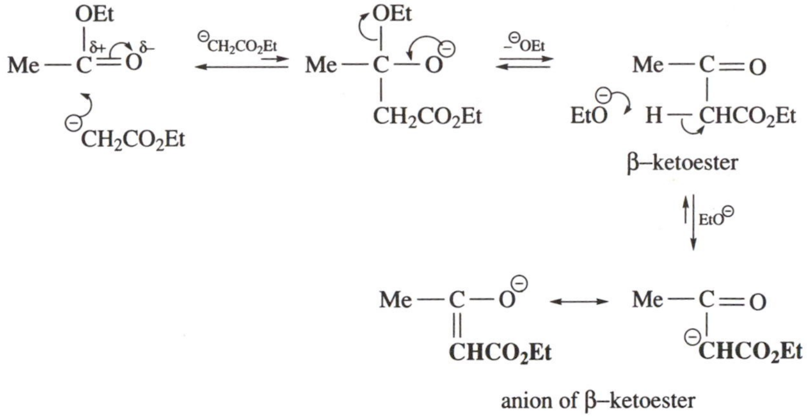

[7.32] ADDITION OF $^{\ominus}CH_2CO_2Et$ TO CH_3CO_2Et
(CLAISEN ESTER CONDENSATION)

This reaction differs from the similar ones with aldehydes and ketones, in that the product from initial carbanion addition has an EtO group attached to what was the carbonyl carbon atom in the original ester molecule. EtO is an excellent leaving group (*cf.* **2.1.7**, p. 23)—in the form of the stable EtO$^{\ominus}$—and is readily lost to form the product, a β-ketoester. The equilibrium yield of product is poor, but this can be overcome by using slightly more than a mole of NaOEt which, by converting the β-ketoester into its stable (delocalised) anion, pulls the equilibrium across to the right-hand side in favour of the product. This reaction is known as the **Claisen** ester condensation.

A useful special case of this reaction is when the acidic CH_2 group, and the C=O group to which addition takes place, are both parts of the same molecule:

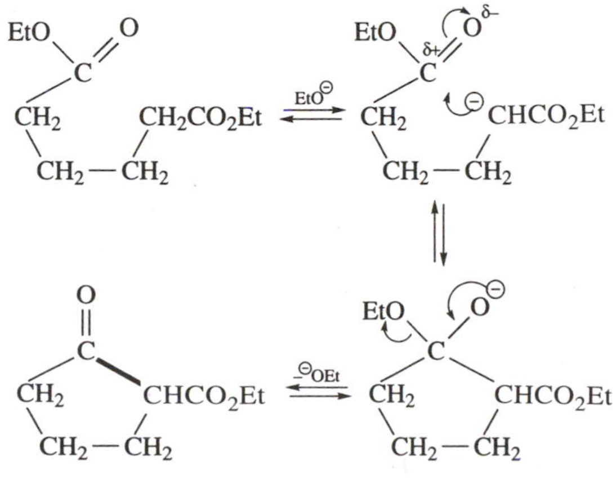

[7.33] DIECKMANN REACTION ON $EtO_2C(CH_2)_4CO_2Et$

The result is a useful synthesis of cyclic compounds, and the intramolecular Claisen reaction is known as a **Dieckmann** cyclisation.

7.2.7   Addition of nitrogen nucleophiles

Nitrogen nucleophiles will also add to the C=O group in aldehydes and ketones, as we should expect (*cf.* [2.23], p. 23), but the reaction does not normally stop at simple addition:

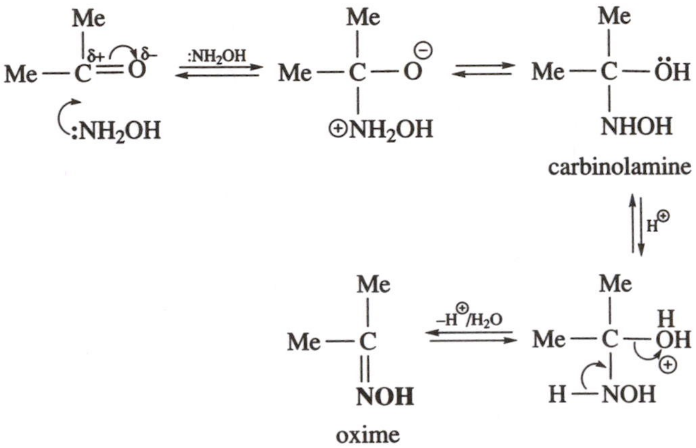

[7.34] ADDITION OF X—NH$_2$ NUCLEOPHILES TO C=O

Thus in the reaction of NH$_2$OH (hydroxylamine) with propanone, the product of simple addition—a **carbinolamine**—readily loses H$_2$O to form the ultimate product, an **oxime**, which contains a C=N bond.

The reaction between NH$_2$OH and a carbonyl compound is influenced considerably by the pH at which it is carried out, each reaction proceeding most readily at a slightly different (optimum) pH; these optimum values are, however, all found to lie in the very narrow band, pH 4–5. This is due to the need to strike a balance between the acid-catalysis that is required to promote dehydration of the carbinolamine intermediate and avoidance of protonation of :NH$_2$OH, which would prevent it acting as a nucleophile towards the C=O group in the first place.

The formation of the carbinolamine intermediate may be observed spectroscopically and, when there are powerfully electron-withdrawing substituents attached to the original carbonyl C=O, this intermediate may be sufficiently stabilised to allow its actual isolation (*cf.* stabilisation of carbonyl hydrates in [7.10], p. 108). Both hydroxylamine and phenylhydrazine (C$_6$H$_5$NHNH$_2$: especially its 2,4-dinitro derivative) are classical reagents for converting liquid aldehydes and ketones into solid derivatives.

7.3   ADDITION TO RCOX

Nucleophiles will also add to the C=O group in derivatives of carboxylic acids such as RCO$_2$Et (esters), RCOCl (acid chlorides), RCOOCOR (acid anhydrides), etc. They do not add to the C=O group in carboxylic acids themselves, however, because an easier reaction for the electron-rich nucleophile is to act as a base, and simply remove a proton, e.g. with $^{\ominus}$OH:

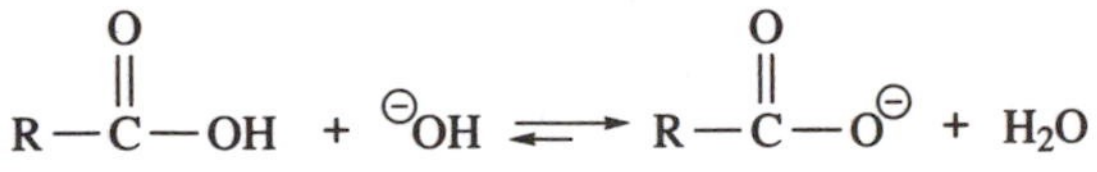

$$R-\overset{\overset{\displaystyle O}{\|}}{C}-OH \ + \ ^{\ominus}OH \ \rightleftharpoons \ R-\overset{\overset{\displaystyle O}{\|}}{C}-O^{\ominus} \ + \ H_2O$$

[7.35] REACTION OF RCO_2H WITH $^{\ominus}$OH

The resultant carboxylate anion, $RCO_2^{\ominus}$, is now itself electron-rich and so will tend to repel nucleophiles, but more importantly this delocalised anion no longer contains an actual C=O group for a nucleophile to add to:

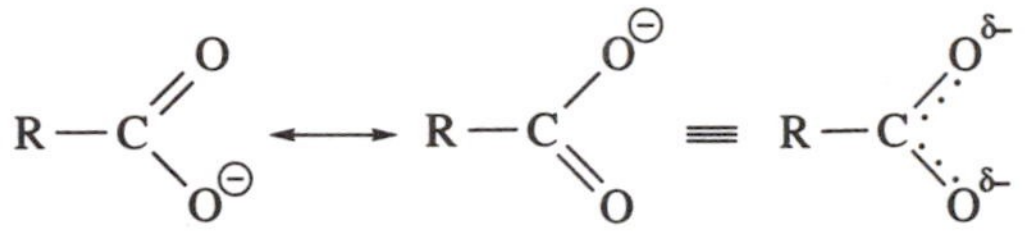

[7.36] NO C=O IN DELOCALISED CARBOXYLATE ANION, $RCO_2^{\ominus}$

In general the sequence of reactivity of RCOX towards nucleophiles is found to be that shown in [7.37]:

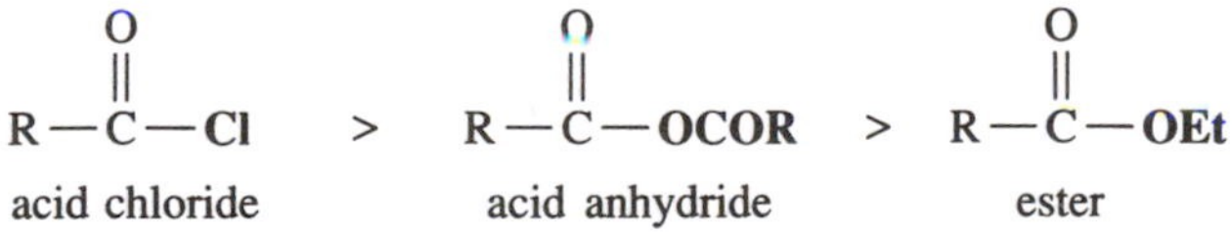

$$R-\overset{\overset{\displaystyle O}{\|}}{C}-Cl \quad > \quad R-\overset{\overset{\displaystyle O}{\|}}{C}-OCOR \quad > \quad R-\overset{\overset{\displaystyle O}{\|}}{C}-OEt$$

acid chloride   acid anhydride   ester

[7.37] SEQUENCE OF REACTIVITY IN RCOX TOWARDS
NUCLEOPHILES

Thus many acid chlorides are sufficiently reactive for them to be attacked vigorously even by nucleophiles as weak as H_2O:, not requiring the more powerful $^{\ominus}$OH:

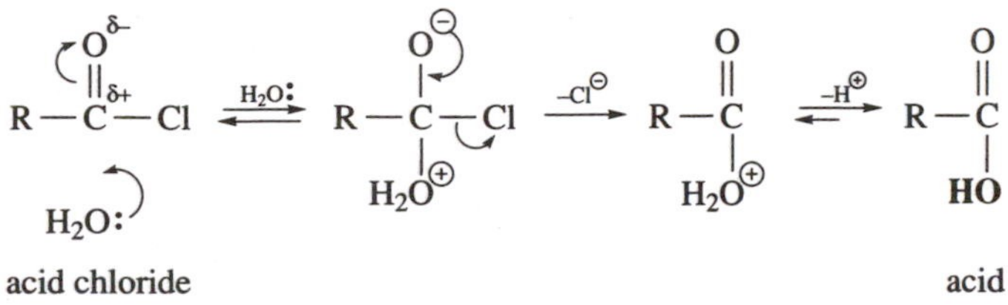

acid chloride   acid

[7.38] HYDROLYSIS OF RCOCl WITH H_2O

The reaction does not stop at simple addition, however, but involves the subsequent loss of a group from the initial addition product. This is because the original carbonyl carbon atom in RCOX (unlike that in aldehydes and ketones) carries a substituent, X, which is a good potential leaving group, e.g. $Cl^\ominus$ in [7.38] is a particularly good one. The reactivity sequence for RCOX listed in [7.37] (p. 121) reflects, in part at least, the relative ability as leaving groups of $Cl^\ominus$, $RCO_2^\ominus$, and $EtO^\ominus$.

While some acid chlorides are reactive enough to undergo hydrolysis with H_2O (*cf.* [7.38], p. 121), esters normally require $^\ominus OH$:

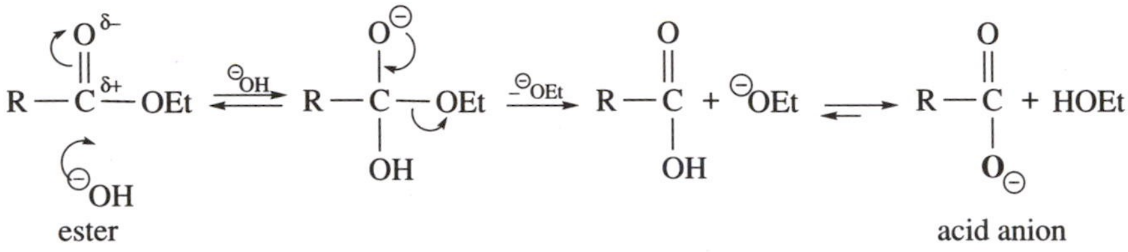

[7.39] ESTER HYDROLYSIS WITH $^\ominus OH$

The final products are the carboxylate anion, $RCO_2^\ominus$, and the alcohol, EtOH; this reaction is not reversible, because any reverse attack by $^\ominus OEt$ on RCO_2H would lead, as it does in [7.39], merely to loss of a proton from its CO_2H group.

The hydrolysis of esters may also be carried out under acid-catalysed conditions. Here initial addition is of the weaker nucleophile H_2O:, but this is made possible by protonation of the oxygen atom of the C=O group in the ester, thereby making the carbonyl carbon atom more susceptible to attack by the weaker nucleophile:

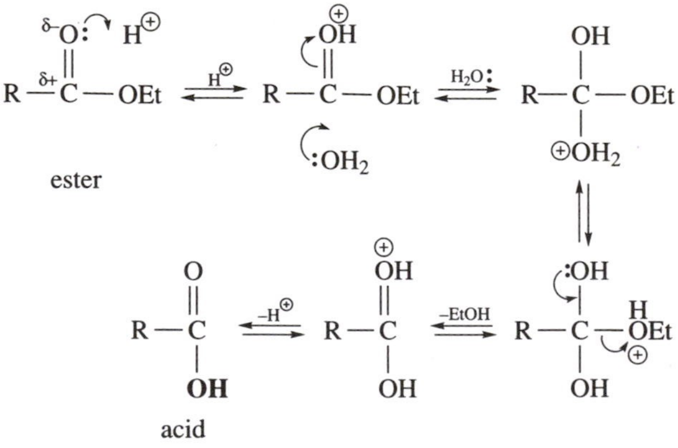

[7.40] ACID-CATALYSED ESTER HYDROLYSIS

Under these conditions the reaction is completely reversible: for hydrolysis, the excess of H_2O in dilute aqueous acid will drive the equilibrium over to the right-hand side; while for ester-formation, a non-aqueous acid catalyst, e.g. concentrated H_2SO_4, in an excess of the alcohol (in this case EtOH) will drive the equilibrium over to the left-hand side. There is more detailed discussion of acid-catalysed ester hydrolysis in **9.3.2** (p. 148).

The reason that similar reactions—involving loss of a group subsequent to initial addition—do not take place, when nucleophiles are added to aldehydes and ketones, is that the species that would have to be lost—$H^{\ominus}$ or $R^{\ominus}$ from aldehydes, and $R^{\ominus}$ from ketones—are such extremely poor leaving groups. If, however, electron-withdrawing substituents are introduced into such an R group, these will serve to stabilise R as $R^{\ominus}$ and thereby improve its ability as a potential leaving group; reactions involving loss of such a modified leaving group, subsequent to initial nucleophilic addition, are indeed observed, e.g. the haloform reaction; so named because the product, with a CCl_3 substituent, is chloroform $HCCl_3$:

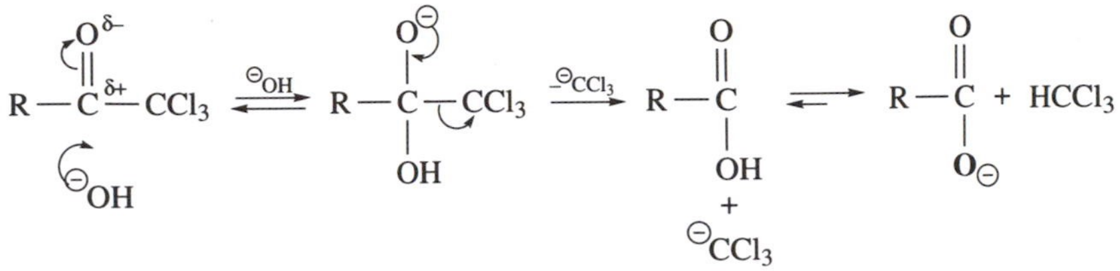

[7.41] HALOFORM REACTION: $^{\ominus}CCl_3$ AS LEAVING GROUP

7.4 ADDITION TO C≡N

The C≡N triple bond is polarised in the same way as C=O, because nitrogen too is more electronegative than carbon, and will thus draw the electrons in the bond towards itself: nucleophilic addition to C≡N would therefore be expected to take place. Nitriles, RC≡N, are indeed found to undergo hydration, under both acid- and base-catalysed conditions:

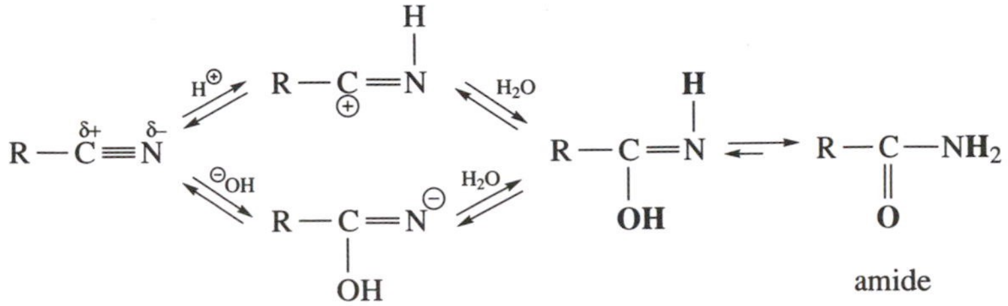

[7.42] HYDRATION OF RC≡N

The product of initial addition is an **amide**, $RCONH_2$, but it is often difficult to prevent the reaction proceeding further to yield the acid anion, $RCO_2^{\ominus}$, when aqueous base is the reagent, or the acid itself, RCO_2H, when aqueous acid is used.

7.5 SUMMARY

Nucleophilic addition to the C=C bond requires the presence of an electron-withdrawing substituent to induce $+^{ve}$ polarisation in the unsubstituted carbon atom of the bond; e.g. C=C—CN, where nucleophiles add readily to the unsubstituted carbon in a useful synthetic procedure: **cyanoethylation.** HBr can, similarly, be added to C=C—C=O; but potentially more useful is the addition of organo-metallic reagents, e.g. RMgBr, to its C=C bond (as well as the more familiar addition to C=O).

Much more important, however, is nucleophilic addition to the C=O bond, which is itself polarised, $^{\delta+}C=O^{\delta-}$; such addition is found to be influenced considerably by the steric and electronic effects that may be exerted by the groups attached to the carbonyl carbon atom. The addition of a number of different nucleophiles, e.g. H_2O, ROH, RSH, HCN and H_2 (*via* metal hydride ions) is then considered in detail. Most of these reactions are reversible, except that of $H^{\ominus}$ which is essentially irreversible.

More significant—particularly in synthetic terms—is the addition of carbon nucleophiles to C=O. These include organo-metallic compounds, e.g. RMgBr, acetylide anions, $RC\equiv C^{\ominus}$, and, most importantly, the carbanions obtained by loss of proton from other carbonyl compounds. Thus aldehydes, RCHO, can lead to the **aldol** reaction, and also to hydride transfer in the **Cannizzaro** reaction. Some ketones will undergo an aldol type reaction, while esters—which contain OR as a potential leaving group—can undergo the **Claisen** ester condensation, and the **Dieckmann** reaction (cyclisation).

The addition of nitrogen nucleophiles, e.g. hydroxylamine, NH_2OH, is then considered; the initial carbinolamine intermediate here losing H_2O to form an oximino group, C=NOH.

Initial nucleophilic addition to the C=O group of carboxylic acid derivatives, RCOX, e.g. esters, acid chlorides, acid anhydrides, etc., is followed by loss of a leaving group from the initial tetrahedral intermediate, e.g. in the hydrolysis of esters. Similar loss of a leaving group, $R^{\ominus}$, from the initial addition product of aldehydes and ketones can also occur if suitable stabilising features are present in R, e.g. the **haloform** reaction.

Finally, nucleophilic addition to C≡N is mentioned; particularly addition of H_2O (hydration), leading to overall hydrolysis.

Elimination

8

Nucleophilic (base-induced) elimination

Elimination reactions involve the loss from a molecule of atoms or groups which are not then replaced by other atoms or groups. The atoms from which such loss takes place are often—though by no means universally—carbon; in the most familiar case, two carbon atoms are involved which are adjacent to each other, and elimination then results in the formation of a double bond between them. Elimination often involves the loss of H from one of these carbon atoms, and may then be induced by electron-rich species (e.g. $^{\ominus}$OH), which are acting here as **bases**, rather than as nucleophiles (*cf.* **2.1**, p. 13).

8.1 ELIMINATION TO FORM C=C

A good example is the overall elimination of HY from $H—CH_2CH_2—Y$ (e.g. $Y = Br$ in CH_3CH_2Br), induced by $^{\ominus}$OH, to form $CH_2\text{=}CH_2$:

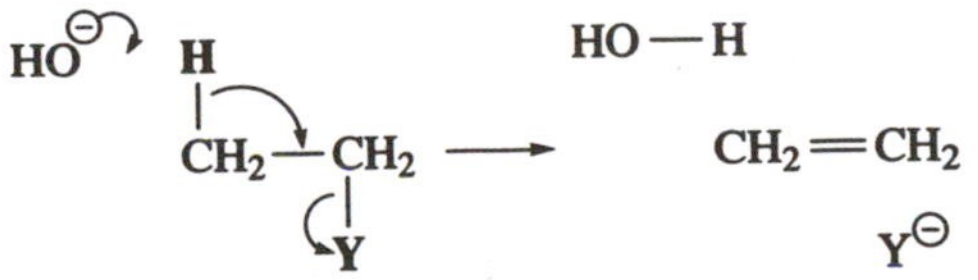

[8.1] $^{\ominus}$OH INDUCED ELIMINATION OF HY FROM $H—CH_2CH_2—Y$

As loss of H and Y takes place from adjacent (1,2- or αβ-) atoms, this type of reaction is called 1,2- or αβ-(usually just β-) elimination. We have already seen a hydroxyl ion, $^{\ominus}$OH, acting as a nucleophile towards C—Br in simple alkyl bromides (**2.1**, p. 13), here it is acting as a base towards H—C in H—CH$_2$CH$_2$Y.

The salient features, in the actual pathway by which this elimination reaction may proceed, are the breaking of H—C and C—Y bonds. We can imagine this happening in any one of three different sequences: (1) a single step, concerted pathway in which both H—C and C—Y bonds are broken simultaneously (**E2** pathway); (2) a two step pathway in which the C—Y bond is broken first, followed by the H—C bond (**E1** pathway); and (3) a two step pathway in which the H—C bond is broken first, followed by the C—Y bond (**E1cB** pathway). Examples are known of 1,2-elimination reactions, to form C=C, that proceed *via* each of these three modes, but the **E1cB** pathway (**8.1.3**, p. 131) is much the most uncommon.

8.1.1   E2 reaction pathway

This one step pathway for elimination is essentially the equivalent of the S$_N$2 pathway for nucleophilic substitution at a saturated carbon atom (**2.1.2**, p. 14); here **E2** is describing an Elimination reaction in which **2** species appear in the rate equation. Thus kinetic experiments establish the rate equation for elimination of HBr from CH$_3$CH$_2$Br, induced by $^{\ominus}$OH, to be:

$$\text{RATE} = k\ [\text{H} - \text{CH}_2\text{CH}_2 - \text{Br}]\ [^{\ominus}\text{OH}]$$

[8.2] RATE EQUATION FOR ELIMINATION OF HBr FROM
H—CH$_2$CH$_2$—Br INDUCED BY $^{\ominus}$OH

The simplest rationalisation for such a rate equation would (as with the S$_N$2 pathway, **2.1.2**, p. 14) be a simple collision between the two species involved: a one step pathway in which both H—C and C—Y bonds are broken simultaneously:

[8.3] E2 PATHWAY FOR ELIMINATION OF HBr FROM
H—CH$_2$CH$_2$—Br

This does, of course, raise the question of what evidence there is that this reaction does indeed proceed *via* a single step that involves breaking both the relevant bonds? Perhaps most cogently, much seeking has failed to detect any sign of an intermediate, which would necessarily be involved in any two step pathway. Evidence that the H—C bond is broken in the vital step of the reaction is provided by replacing the H atoms in the CH_3 group of CH_3CH_2Br by the heavier isotope deuterium, **D**:

$$HO^{\ominus} \quad \mathbf{D} \qquad\qquad HO—D$$
$$\underset{\underset{Br}{|}}{CD_2}—CH_2 \longrightarrow CD_2{=}CH_2$$
$$Br^{\ominus}$$

[8.4] KINETIC ISOTOPE EFFECT IN ELIMINATION OF **D**—Br FROM **D**—CD_2CH_2—Br

The rate of $^{\ominus}OH$ induced elimination from **D**—CD_2CH_2Br is then found to be markedly slower than elimination from H—CH_2CH_2Br, under the same conditions. There is a **kinetic isotope effect** (*cf.* **3.2.1.3**, p. 34), which indicates that H—C, and **D**—C, bond-breaking must be taking place in the step whose rate our kinetic experiments are measuring.

Similarly, the rate of overall elimination is influenced—hardly surprisingly—if Y, in CH_3CH_2—Y, is changed from Br to some other leaving group (e.g. Cl); so the C—Y bond, also, must be broken in the step whose rate our kinetic experiments are measuring. **E2** is found to be the commonest, and most straightforward, of the pathways for elimination, but before more detailed discussion of it reference will be made to the other pathways for 1,2-elimination: **E1** and **E1cB**.

8.1.2   E1 reaction pathway

This two step pathway for elimination, in which the C—Y bond is broken before the H—C bond, is essentially the equivalent of the S_N1 pathway for nucleophilic <u>substitution</u> at a saturated carbon atom (**2.1.3**, p. 15); here **E1** is describing an <u>E</u>limination reaction in which only <u>**1**</u> species appears in the rate equation. Thus kinetic experiments establish the rate equation for elimination of HBr from $MeCH_2$—CMe_2Br, induced by $^{\ominus}OH$, to be:

$$RATE = k \ [MeCH_2 — CMe_2Br]$$

[8.5] RATE EQUATION FOR ELIMINATION OF HBr FROM $MeCH_2$—CMe_2Br INDUCED BY $^{\ominus}OH$

This demonstrates that the vital step that controls the rate of overall elimination cannot involve $^{\ominus}OH$, and that this step must therefore be a "do-it-yourself" operation on the part of the alkyl bromide alone. This step, as with S_N1 (2.1.3, p. 15), is most likely to be **ionisation** to form $Br^\ominus$ and a **carbocation** intermediate:

$$\underset{\underset{Br}{|}}{\overset{\overset{H}{|}}{MeCH}}-CMe_2 \xrightarrow{\text{slow}} \underset{\underset{\oplus}{|}}{\overset{\overset{HO^\ominus \; H}{|}}{MeCH}}-CMe_2 \quad Br^\ominus \xrightarrow{\text{fast}} MeCH=CMe_2$$

$$\text{carbocation intermediate}$$

[8.6] E1 PATHWAY FOR ELIMINATION OF HBr FROM
$MeCH_2—CMe_2Br$

This slow, rate-limiting step is then followed by rapid removal of $H^\oplus$ from the carbocation intermediate by $^\ominus OH$ to complete the overall elimination of HBr.

This carbocation intermediate is, of course, exactly the same species that we encountered in S_N1 substitution reactions of alkyl bromides with $^\ominus OH$. In S_N1 reactions $^\ominus OH$ then acts as a nucleophile towards the carbocation intermediate leading to overall <u>substitution</u>, while in E1 reactions $^\ominus OH$ then acts as a <u>base</u> towards the carbocation leading to overall <u>elimination</u>. Hardly surprisingly, therefore, it is not uncommon in such reactions to obtain a mixture of elimination and substitution products. There will, therefore, be a general discussion of the factors that influence **elimination** versus **substitution** below (**8.1.6**, p. 135).

There is also the question of what encourages an elimination reaction to proceed *via* the E1 pathway rather than *via* the E2. Clearly the E1 pathway will be encouraged by any feature that promotes ionisation of the C—Y bond. These include: (1) Y being a good <u>leaving group</u> (as $Y^\ominus$ or Y:); (2) carrying the reaction out in a solvent that assists ionisation, and that also serves to stabilise the developing ions through solvating them; and (3) structural features in the R group of R—Y that serve to stabilise the intermediate, $R^\oplus$, as a carbocation.

As we saw in [5.19] (p. 76), the relative stability of carbocations follows the order:

$$\overset{\oplus}{R_3C} \; > \; \overset{\oplus}{R_2CH} \; > \; \overset{\oplus}{RCH_2} \; > \; \overset{\oplus}{CH_3}$$

[8.7] RELATIVE CARBOCATION STABILITY

Elimination *via* the E1 pathway should thus be promoted progressively on going from primary to secondary to tertiary halides (e.g. Y = Br); which is exactly what we find to occur:

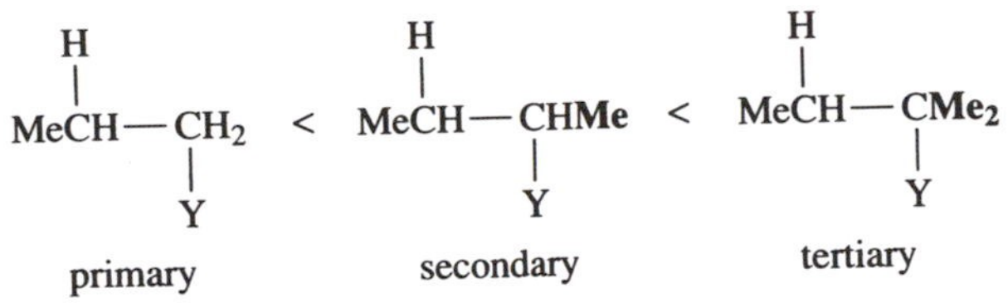

$$
\underset{\text{primary}}{\overset{\overset{\displaystyle H}{|}}{\text{MeCH}-\underset{\underset{\displaystyle Y}{|}}{\text{CH}_2}}} \quad < \quad \underset{\text{secondary}}{\overset{\overset{\displaystyle H}{|}}{\text{MeCH}-\underset{\underset{\displaystyle Y}{|}}{\text{CHMe}}}} \quad < \quad \underset{\text{tertiary}}{\overset{\overset{\displaystyle H}{|}}{\text{MeCH}-\underset{\underset{\displaystyle Y}{|}}{\text{CMe}_2}}}
$$

[8.8] EFFECT OF STRUCTURE ON PROMOTION OF E1 PATHWAY
FOR ELIMINATION

8.1.3 ElcB reaction pathway

This two step pathway for elimination, in which the H—C bond is broken before the C—Y bond, is found to be relatively uncommon for simple 1,2-elimination reactions. Here **ElcB** is describing an **E**limination reaction that proceeds *via* the **c**onjugate **B**ase of the starting material (i.e. the **carbanion** formed from it by loss of proton) as an intermediate:

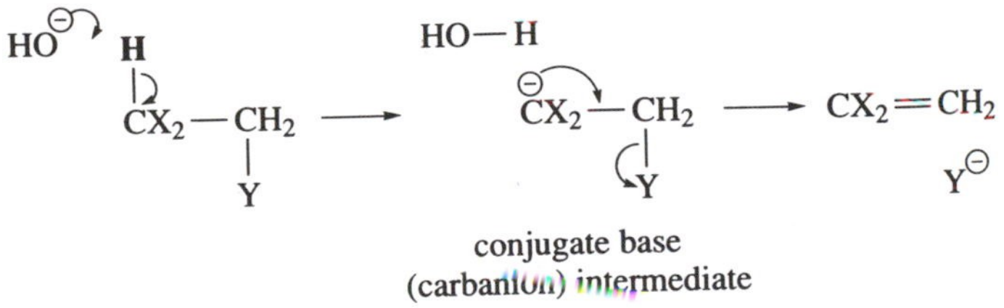

conjugate base
(carbanion) intermediate

[8.9] ElcB PATHWAY FOR ELIMINATION *VIA* A CONJUGATE BASE
(CARBANION) INTERMEDIATE

It might be expected that this pathway would be promoted by any structural features (e.g. substituents) on the β-carbon atom: (1) that serve to increase the acidity of H atoms on this carbon atom, thereby promoting their removal, as $H^{\oplus}$, by base; and (2) that serve to stabilise the carbanion intermediate that results from such loss. The most likely substituents will be electron-withdrawing atoms or groups, but such electron-withdrawal needs, apparently, to be extremely powerful to shift an elimination from an E2 to an **ElcB** pathway. The presence of **two** Cl substitutents on the β-carbon atom is, however, found to lead to elimination of HF from $H-CCl_2CF_2-F$ proceeding *via* the **ElcB** pathway:

carbanion intermediate

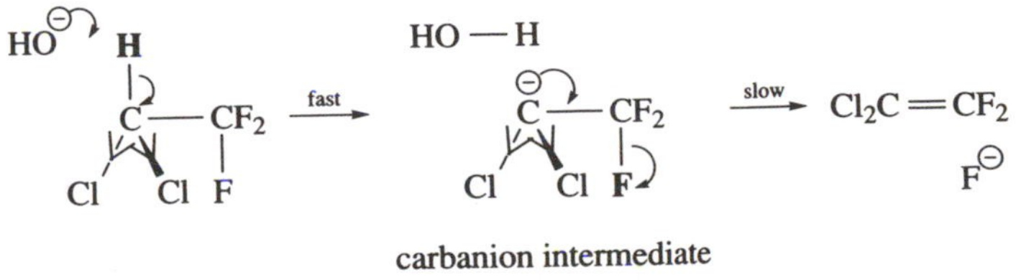

[8.10] ElcB PATHWAY FOR ELIMINATION PROMOTED BY TWO Cl
ATOMS ON THE β-CARBON ATOM

This example illustrates another feature that promotes elimination *via* an **E1cB** pathway, namely a poor leaving group on the α-carbon atom—in this case $F^\ominus$, which is poor indeed. All these features would serve to make breaking of the H—C bond faster than breaking of the C—Y bond.

Establishing whether a particular elimination reaction proceeds *via* an **E1cB** or an **E2** pathway presents some difficulty as the rate equation is commonly the same for both:

$$\text{RATE} = k\ [\text{H} - \text{CX}_2\text{CH}_2 - \text{Y}]\ [^\ominus\text{OH}]$$

[8.11] RATE EQUATION FOR E1cB—AND E2—PATHWAYS FOR ELIMINATION

A distinction between these two pathways can, however, in some cases be made by means other than simple kinetic measurements. A good example is the elimination of HF from H—CCl$_2$CF$_2$—F with $^\ominus$OEt (as base), when this is carried out in EtOD—ethanol in which the hydrogen atom of the OH group has been replaced by the heavier isotope deuterium, **D**:

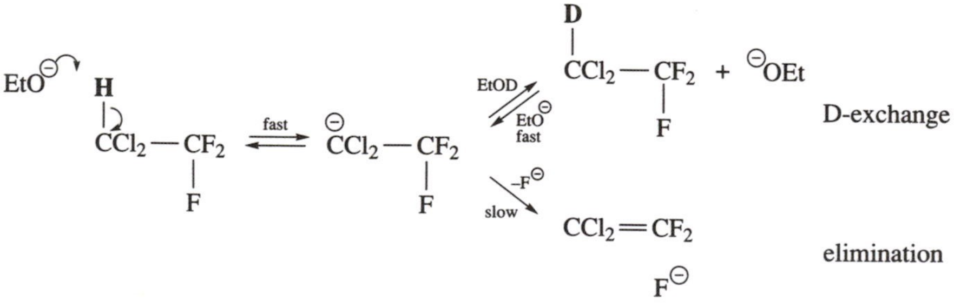

[8.12] H/D EXCHANGE IN E1cB PATHWAY FOR ELIMINATION OF HF FROM H—CCl$_2$CF$_2$—F

If the reaction is stopped well before elimination is complete, and the as yet un-reacted starting material recovered, it is found that its H atom has been largely replaced by **D**. This indicates that the H atom in the original halide is undergoing reversible, base-induced exchange with the **D** atom of the solvent, and that this (reversible) proton removal must, therefore, be faster than a subsequent step in which loss of $F^\ominus$ completes the overall elimination of HF.

Other reactions that are believed to proceed *via* the relatively uncommon **E1cB** pathway include the dehydration of aldols (**7.2.6.3.1**, p. 116) induced— unusually for dehydration (*cf.* **9.1.1**, p. 142)—by a <u>base</u>:

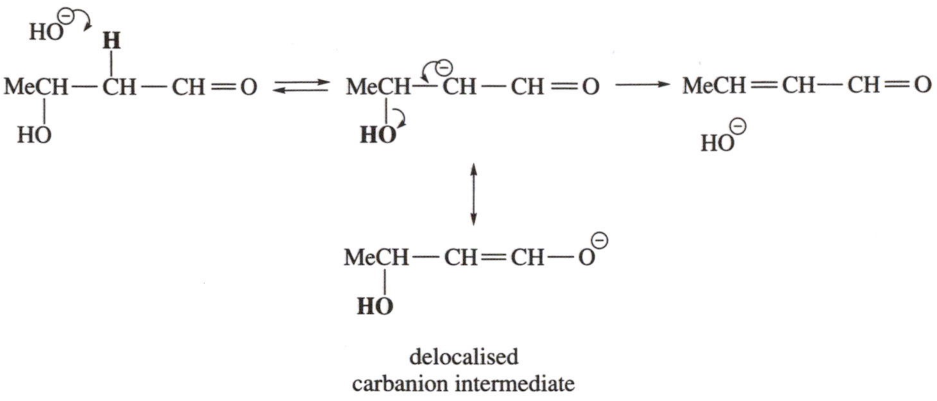

$$\text{MeCH}-\overset{\overset{\displaystyle \text{HO}^{\ominus}\,\text{H}}{|}}{\underset{\underset{\displaystyle \text{HO}}{|}}{\text{CH}}}-\text{CH}=\text{O} \rightleftharpoons \text{MeCH}-\overset{\overset{\ominus}{}}{\underset{\underset{\displaystyle \text{HO}}{|}}{\text{CH}}}-\text{CH}=\text{O} \longrightarrow \text{MeCH}=\text{CH}-\text{CH}=\text{O}$$

$$\text{HO}^{\ominus}$$

$$\updownarrow$$

$$\text{MeCH}-\underset{\underset{\displaystyle \text{HO}}{|}}{\text{CH}}=\text{CH}-\text{O}^{\ominus}$$

delocalised
carbanion intermediate

[8.13] E1cB BASE-INDUCED DEHYDRATION OF ALDOLS

Here the acidity of the H atom is promoted by the adjacent electron-withdrawing C=O group, which also stabilises the developing carbanion intermediate through delocalisation of its negative charge; in addition, $^{\ominus}$OH is not a particularly good leaving group.

Another example of an unusual **E1cB** elimination is the formation of aryne intermediates (*cf*. **2.2.2**, p. 27), particularly when the leaving group is $F^{\ominus}$:

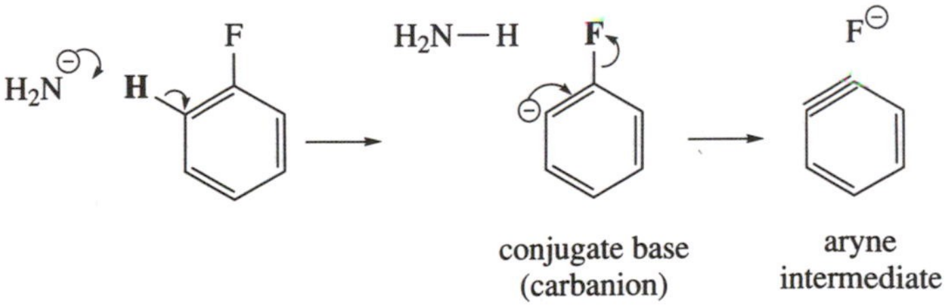

conjugate base   aryne
(carbanion)   intermediate

[8.14] E1cB PATHWAY FOR FORMATION OF ARYNE
INTERMEDIATES

8.1.4 Stereochemistry of elimination

In a molecule such as $H-CH_2-CH_2-Y$ there is, of course, unrestricted rotation about the bond connecting the α- and β-carbon atoms, so that H and Y can take up an essentially infinite number of different positions relative to each other—these different, but interconvertible, relative positions of atoms within the molecule are called **conformations**. It does, however, seem not unreasonable to suppose that one or more of all these different possible conformations would offer some advantage, in facilitating the elimination of H and Y in the single concerted step of an **E2** elimination pathway.

In just two of these conformations, H and Y are both in the same plane; such conformations are described as being *periplanar*:

syn-periplanar
conformation

anti-periplanar
conformation

[8.15] SYN- AND ANTI-PERIPLANAR CONFORMATIONS OF
$H—CR_2CR_2—Y$

In one of these two conformations H and Y are both on the <u>same</u> side of the molecule (*syn-periplanar* conformation), while in the other they are on <u>opposite</u> sides (**anti**-*periplanar* conformation). There is reason to believe that formation of the developing double bond is assisted when **E2** elimination takes place from those conformations in which H and Y are *periplanar*; there remains, however, the question of whether elimination occurs more readily from the *syn-*, or from the *anti-*, conformation.

This point can readily be put to experimental test by observing the structure of the alkene obtained by elimination of HBr from the bromide, **H—CRR′CRR′—Br**, in which both α- and β-carbon atoms carry non-identical substitutents, R and R′:

cis alkene

syn-periplanar
conformation

anti-periplanar
conformation

trans alkene

[8.16] ELIMINATION OF HBr FROM SYN- AND ANTI-PERIPLANAR
CONFORMATIONS OF H—CRR′CRR′—Br

Elimination from the *syn-periplanar* conformation would lead to formation of the <u>cis</u> alkene, in which similar alkyl groups, e.g. R and R (on the adjacent carbon atoms), are on the <u>same</u> side of the double bond. Conversely, elimination from the *anti-periplanar* conformation would lead to formation of the *trans* alkene, in which similar alkyl groups, e.g. R and R (on the adjacent carbon atoms), are on <u>opposite</u> sides of the double bond. In simple cases such as this one, it is in fact the *trans* alkene that is obtained almost exclusively.

This preference for elimination from the *anti-periplanar* conformation seems not unreasonable when we consider that the attacking base, $^{\ominus}$OH—as it approaches the H atom to be eliminated—is as far away from the very bulky Br atom as it possibly can be. Furthermore, the electron pair that is developing on the β-carbon atom, as its proton is removed by $^{\ominus}$OH, is then able to attack the α-carbon atom from the <u>back</u>, as the negatively charged leaving group, Br$^{\ominus}$, (with its electron cloud) departs from the opposite side. This bears some resemblance to attack by the electron pair of a nucleophile on such a carbon atom in the S_N2 pathway for substitution (**2.1**, p. 13).

This degree of **stereoselectivity** in elimination is commonly observed only for the one step **E2** pathway. Both the two step pathways, **E1** and **E1cB**, proceed *via* intermediates (carbocations [8.6] (p. 130) and carbanions [8.9] (p. 131), respectively) in which the arrangement of groups about the relevant carbon atoms is planar; completion of overall elimination can thus lead to the formation of either *cis* or *trans* alkene, or most commonly to a mixture of both.

8.1.5 Elimination of groups other than HHal

We have to date considered almost exclusively the elimination of HBr to form C=C, but there are many other 1,2-elimination reactions in which atoms or groups other than Br$^{\ominus}$ are lost (as well as H). Thus loss of R_3N: from H—CH_2CH_2—$NR_3^{\oplus}$ is found to occur readily, reflecting the effectiveness of R_3N: as a leaving group. There are also many examples of 1,2-elimination reactions which do not involve H as one of the atoms eliminated. Thus the elimination of Br_2 from 1,2-dibromides may be promoted by the iodide ion, $I^{\ominus}$, or by Zn metal; though this is seldom of preparative value as the 1,2-dibromide was almost certainly made by addition of Br_2 to the alkene in the first place!

A particularly interesting example is the elimination of Br$^{\ominus}$ and CO_2 from the acid PhCHBr—CHBrCO$_2$H (obtained by addition of Br_2 to the unsaturated acid, PhCH=CHCO$_2$H: 3-phenylprop-2-enoic acid, i.e. cinnamic acid):

[8.17] ELIMINATION OF Br$^{\ominus}$ AND CO_2 FROM PhCHBr—CHBrCO$_2$H

The elimination proceeds under extremely mild conditions: merely dissolving the acid in aqueous base! This remarkable ease of elimination no doubt reflects just how good a molecule of CO_2 is as a leaving group.

8.1.6 Elimination versus substitution

We have already mentioned that, for a given alkyl halide, the two step **E1** pathway for elimination proceeds *via* the same carbocation intermediate as the two step S_N1 pathway for nucleophilic substitution:

$$
\begin{array}{c}
\text{H} \\
| \\
\text{MeCH—CMe}_2 \\
| \\
\text{Br}
\end{array}
\xrightarrow{\text{slow}}
\begin{array}{c}
\text{H} \\
| \\
\text{MeCH—}\overset{\oplus}{\text{CMe}}_2 \\
\\
\text{Br}^{\ominus}
\end{array}
$$

HO—H

MeCH=CMe$_2$ elimination

$\overset{\ominus}{\text{OH}}$ ↗ E1
fast

$\searrow$ S_N1
$\overset{\ominus}{\text{OH}}$ ↖

H
|
MeCH—CMe$_2$
|
OH

substitution

[8.18] E1 *VERSUS* S$_N$1 PATHWAYS FOR ELIMINATION/NUCLEOPHILIC SUBSTITUTION

Attack by $^{\ominus}$OH on the carbocation intermediate in the second, fast step of the overall reaction could then lead to the product of elimination (by removal of H$^{\oplus}$), or of substitution (by addition of $^{\ominus}$OH), or more commonly to a mixture of both.

The relationship between the **E2** pathway for elimination and the **S$_N$2** pathway for nucleophilic substitution is not as close, in that each reaction proceeds *via* a single step and these steps are entirely different from each other:

$$
\begin{array}{c}
\text{H}\quad\text{OH} \\
|\quad\ | \\
\text{CH}_2\text{—CH}_2 \\
| \\
\text{Br}^{\ominus}
\end{array}
\xleftarrow[S_N2]{\ominus OH}
\begin{array}{c}
\text{H}\quad\overset{\ominus}{\text{OH}} \\
|\quad\ \\
\text{CH}_2\text{—CH}_2 \\
| \\
\text{Br}
\end{array}
\equiv
\begin{array}{c}
\text{HO}^{\ominus}\ \text{H} \\
| \\
\text{CH}_2\text{—CH}_2 \\
| \\
\text{Br}
\end{array}
\xrightarrow[E2]{\ominus OH}
\begin{array}{c}
\text{HO—H} \\
\\
\text{CH}_2\text{=CH}_2 \\
\\
\text{Br}^{\ominus}
\end{array}
$$

substitution   elimination

[8.19] E2 *VERSUS* S$_N$2 PATHWAYS FOR ELIMINATION/NUCLEOPHILIC SUBSTITUTION

Again, however, there is the possibility of a particular reaction proceeding *via* either or both pathways, leading to the product of elimination, or of substitution, or more commonly to a mixture of both.

Whether a particular reaction yields the product of elimination, or of substitution (or both), is obviously a matter of major preparative importance; we now have to consider, therefore, what factors can influence this outcome, and how it may—to some extent at least—be controlled.

It might be expected that any structural features in the starting material (e.g. substituents on the α- or β-carbon atoms) that serve to stabilise the developing double bond would promote elimination, at the expense of substitution. Alkyl substituents, on the double bond carbon atoms, are known to stabilise alkenes, and the proportion of elimination is indeed found to increase as substitution at the α-carbon atom increases, in the series of bromides in [8.20]:

$$CH_3 - CH_2Br \qquad CH_3 - CHMeBr \qquad CH_3 - CMe_2Br$$

$$\downarrow \qquad\qquad\qquad \downarrow \qquad\qquad\qquad \downarrow$$

$$CH_2 = CH_2 \qquad\quad CH_2 = CHMe \qquad\quad CH_2 = CMe_2$$

primary $\quad < \quad$ secondary $\quad < \quad$ tertiary

[8.20] INCREASING PROPORTION OF ELIMINATION:
PRIMARY < SECONDARY < TERTIARY BROMIDES

A particularly marked example occurs with $C_6H_5CH_2 - CH_2Br$, where the developing double bond would be stabilised, in the course of elimination, through conjugation with the aromatic system of the benzene ring; thus under reaction conditions in which $CH_3 - CH_2Br$ yields only 1% of alkene, $C_6H_5CH_2 - CH_2Br$ is found to yield 99%!

Another significant influence—and one that we <u>can</u> control—is the **size** of the attacking base/nucleophile: the <u>larger</u> this is the greater the proportion of <u>elimination</u> that is found to occur. Thus in the $E1/S_N1$ case, elimination

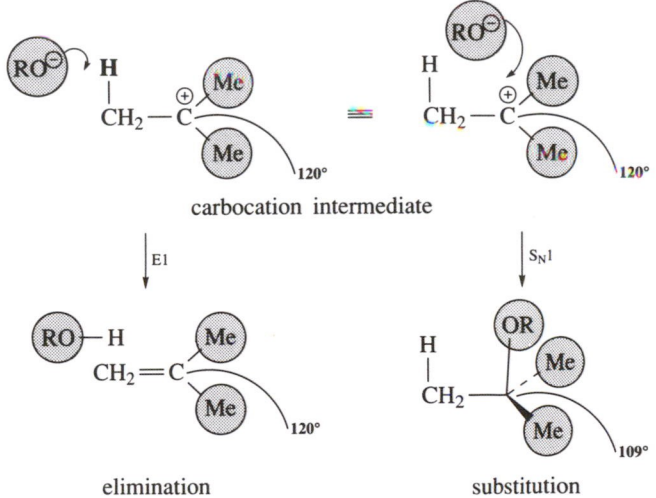

carbocation intermediate

elimination   substitution

[8.21] EFFECT OF SIZE OF BASE/NUCLEOPHILE ($^{\ominus}$OR) ON
$E1/S_N1$ PROPORTION

merely requires the removal of proton from the uncrowded periphery of the carbocation intermediate, whereas substitution requires attack on its crowded, central carbon atom—a much more difficult operation in steric terms.

Substitution may be avoided, and elimination thus promoted, by use of the very bulky base/nucleophile $Me_3CO^{\ominus}$ rather than $CH_3O^{\ominus}$ or $MeCH_2O^{\ominus}$. Tertiary amines, $R_3N:$, also promote elimination, although they are not particularly strong bases, because the bulky R groups attached to the N atom make them—for steric reasons—very much poorer nucleophiles.

There will be an essentially analogous steric effect promoting elimination in the **E2/S$_N$2** situation, though probably not quite so decisively.

There is a further steric influence on the **E1/S$_N$1** situation, and this too may be seen in [8.21] (p. 137). Overall substitution involves a <u>decrease</u> in bond angle at the relevant carbon atom from $\sim 120°$, at the planar carbon atom in the carbocation intermediate, to $\sim 109°$, at the now tetrahedral carbon atom in the substitution product. The attachment of the OR group thus introduces a considerable <u>increase</u> in crowding, because there are <u>three</u> bulky groups attached to this now tetrahedral carbon atom. Overall elimination involves no such <u>decrease</u> in bond angle, and **no** <u>increase</u> in crowding as there are only <u>two</u> bulky groups attached to this carbon atom, which remains planar throughout alkene formation.

8.2 ELIMINATION TO FORM C≡C

It is also possible to eliminate <u>two</u> molecules of HBr from, for example, CH_3—$CHBr_2$ to form a C≡C bond, but this elimination requires the use of stronger bases such as $^{\ominus}NH_2$:

$$HCH_2 - CHBr_2 \xrightarrow{\;^{\ominus}NH_2\;} HC \equiv CH$$

[8.22] ELIMINATION OF 2 HBr FROM CH_3—$CHBr_2$ TO FORM C≡C

Stronger bases are required because elimination of the second molecule of HBr from the intermediate bromoalkene, CH_2=$CHBr$, is found to be considerably more difficult than elimination of the first molecule of HBr from CH_3CHBr_2.

In, for example, CHBr=CHBr the atoms to be eliminated, H and Br, are now held rigidly in a *periplanar* orientation by the double bond:

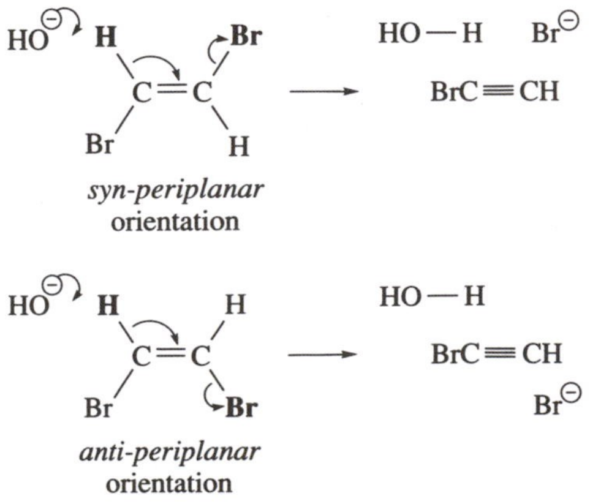

syn-periplanar orientation

anti-periplanar orientation

[8.23] COMPARISON OF ELIMINATIONS IN WHICH H AND Br ARE IN SYN-, AND IN ANTI-, PERIPLANAR ORIENTATIONS

It is, therefore, possible to compare the rate of elimination from the isomer of CHBr=CHBr in which the groups to be eliminated are on the <u>same</u> side of the double bond (*syn-periplanar* isomer), with the rate from the isomer in which they are on <u>opposite</u> sides (*anti-periplanar* isomer). By analogy with what was observed in **8.1.4** (p. 133) for elimination to form C=C, we would expect easier loss of H and Br to occur from the isomer in which these atoms are in an *anti-periplanar* orientation. This is indeed what is observed: the *anti-periplanar* isomer is found to undergo elimination several thousand times faster than the *syn-periplanar* one, under the same conditions.

8.3 ELIMINATION TO FORM C=O

As was mentioned in **7.2.1** (p. 105) many of the nucleophilic addition reactions of the carbonyl group, C=O, are reversible. A typical example is cyanohydrin formation (**7.2.5**, p. 112), whose reversal—to reform the C=O bond of the original carbonyl compound—involves the breaking of a carbon–carbon bond:

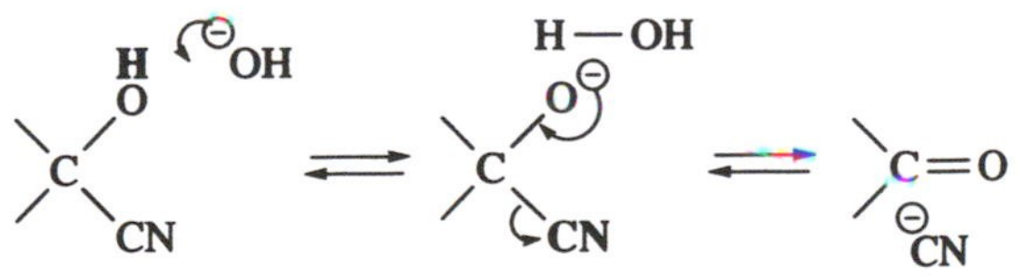

[8.24] REVERSAL OF CYANOHYDRIN FORMATION

This elimination reaction involves rapid, reversible removal of a proton to form an anionic intermediate, which then loses $^{\ominus}$CN in a slower step; it thus proceeds *via* an **E1cB** pathway (**8.1.3**, p. 131). Other examples of elimination reactions to form C=O, through reversal of nucleophilic addition, include reversal of hydration (**7.2.2**, p. 107) and of hemi-acetal formation (**7.2.3**, p. 109).

8.4 1,1-(α) ELIMINATION

To date all the examples that we have considered have been 1,2-elimination reactions in which the two atoms or groups eliminated have been lost from adjacent (1,2-) atoms. Some elimination reactions are, however, known in which both atoms or groups are lost from the <u>same</u> atom, i.e. 1,1-eliminations.

A good example occurs in the hydrolysis of trichloromethane, $CHCl_3$ (chloroform), with strong bases. This reaction is found to follow the rate

$$\text{RATE} = k \text{ [HCCl}_3\text{] [}^{\ominus}\text{OH]}$$

[8.25] RATE EQUATION FOR HCCl$_3$ HYDROLYSIS

equation in [8.25], and is believed to proceed in the following way:

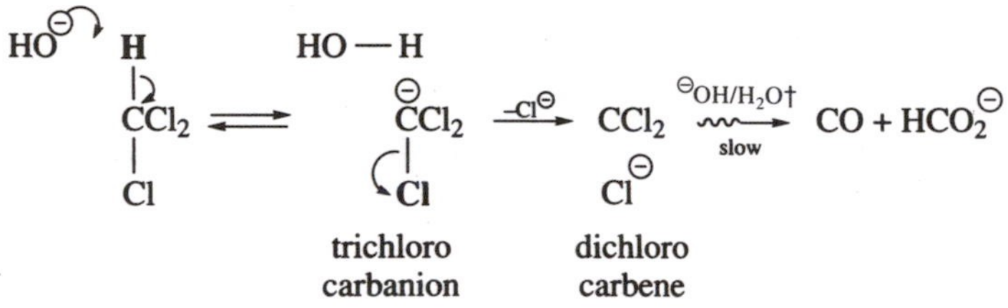

trichloro
carbanion

dichloro
carbene

[8.26] HYDROLYSIS OF HCCl$_3$ WITH STRONG BASE

The hydroxyl ion is here acting as a base and removing a proton from HCCl$_3$ (rather than acting as a nucleophile and displacing Cl$^{\ominus}$) to yield the trichlorocarbanion. This, in turn, can lose Cl$^{\ominus}$ to form the novel intermediate CCl$_2$, dichloro**carbene**, which is then slowly hydrolysed to the observed end products, CO and HCO$_2^{\ominus}$. The 1,1-elimination to form the dichlorocarbene intermediate is a further example of a reaction proceeding *via* an **E1cB** pathway, albeit one involving only a single carbon atom.

By carrying out the reaction in **D$_2$O** (rather than in H$_2$O),

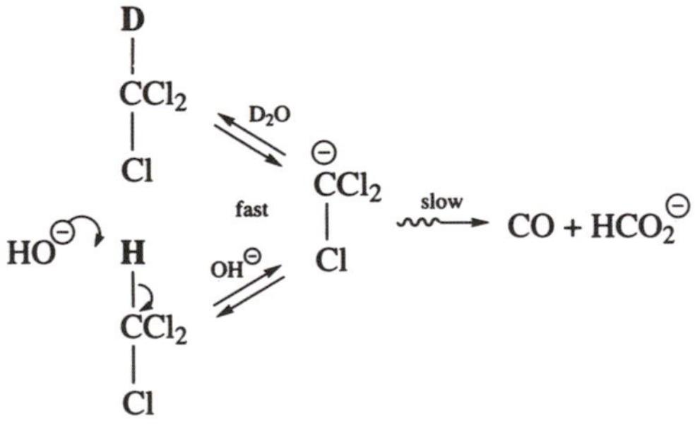

[8.27] HYDROLYSIS OF HCCl$_3$ WITH STRONG BASE IN D$_2$O

and stopping the hydrolysis before it has gone to completion, it can be shown that the as yet unhydrolysed starting material has been converted into **DCCl$_3$**: in other words that loss of proton in [8.26] is very much faster than overall hydrolysis.

There is, however, still the question of the validity of the unusual dichlorocarbene intermediate. The carbon atom in such a species will clearly be highly electron-deficient as it has only <u>six</u> electrons in its outer shell, :CCl$_2$; dichlorocarbene might thus be expected to be highly reactive towards molecules which are themselves electron-rich. We find that if we generate :CCl$_2$ in the presence of an alkene, then ready addition takes place across the double bond to form a three-membered ring, a cyclopropane:

†The symbol ⌇→ indicates that the conversion proceeds *via* several successive steps.

$$\underset{\text{Me}}{\overset{\text{Me}}{\diagdown}} C = C \underset{\text{H}}{\overset{\text{Me}}{\diagup}} \longrightarrow \text{ (cyclopropane product)}$$

Me   Me   Me   Me

RO$^\ominus$ H

$$\underset{\overset{|}{\underset{\text{Cl}}{C}}}{\overset{|}{CCl_2}} \xrightarrow[\text{benzene}]{RO^\ominus} \quad :CCl_2$$

[8.28] ADDITION OF :CCl$_2$ TO MeCH=CHMe

To achieve this addition :CCl$_2$ has to be generated under non-aqueous conditions (in this case in benzene), or it will undergo preferential hydrolysis. Such addition reactions of carbenes in general (:CR$_2$, not merely :CCl$_2$) to alkenes is, indeed, an important method for the preparation of cyclopropanes.

8.5 SUMMARY

Elimination reactions involve the loss from a molecule of atoms or groups, which are not then replaced by other atoms or groups. Much the commonest elimination reactions involve loss of groups from adjacent atoms—usually carbon atoms—and a double bond results ($\alpha\beta$- or 1,2-elimination); one of the atoms or groups lost is very often H. These eliminations are promoted by electron-rich species such as $^\ominus$OH, acting as bases.

Such eliminations are found to proceed *via* any one of three different pathways depending on the sequence in which the bonds to the atoms or groups lost, H and Y, are broken. These comprise: (1) a one step, concerted pathway in which both H—C and C—Y bonds are broken simultaneously (E2 pathway); (2) a two step pathway, *via* a carbocationic intermediate, in which the C—Y bond is broken before the H—C bond (E1 pathway); and (3) a two step pathway, *via* a carbanionic intermediate, in which the H—C bond is broken before the C—Y bond (E1cB pathway).

Consideration is then given to the factors that influence the stereochemistry of 1,2-elimination—particularly to the preferred orientation of H and Y that results in easier elimination; to the elimination of atoms or groups other than H and Br; and to the factors that influence the relative proportions of elimination and substitution observed when the attacking reagent, e.g. $^\ominus$OH, can act as a nucleophile as well as a base.

Other 1,2-elimination reactions leading to the formation of C≡C, and of C=O (reversal of nucleophilic addition to C=O, *cf.* **7.2.1**, p. 105) are then described. Finally there is a consideration of elimination reactions in which both groups are lost from the same atom (1,1-elimination), exemplified by the formation of a dichlorocarbene intermediate during the hydrolysis of HCCl$_3$.

9

Electrophilic (acid-induced) elimination

Electrophilic elimination is very often acid-induced, and involves modifying Y—in, for example, the elimination of HY from HCH_2CH_2Y—by protonation or other means, so as to turn it into a better leaving group. As with base-induced reactions (**8.1**, p. 127), elimination to form C=C is much the most common type.

9.1 ELIMINATION TO FORM C=C

A typical example is the dehydration of alcohols, e.g. $H—CH_2CMe_2—OH$, catalysed by strong acids such as H_2SO_4:

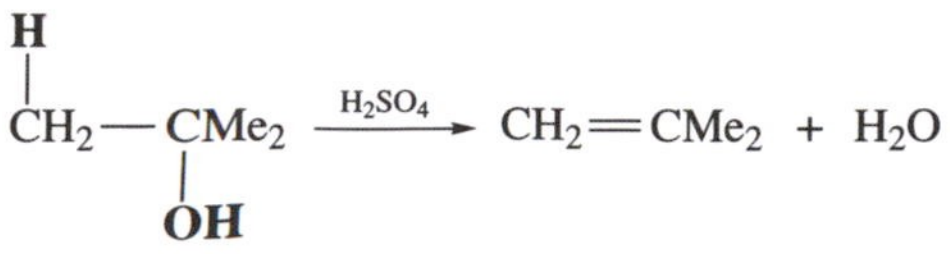

$$\underset{\underset{\displaystyle OH}{|}}{\overset{\overset{\displaystyle H}{|}}{CH_2}}—CMe_2 \xrightarrow{H_2SO_4} CH_2{=}CMe_2 + H_2O$$

[9.1] DEHYDRATION OF $H—CH_2CMe_2—OH$ WITH H_2SO_4

9.1.1 Dehydration of alcohols via protonation of OH

The relative ease with which acid-catalysed dehydration of alcohols takes place is found to follow the sequence in [9.2], which exactly parallels the

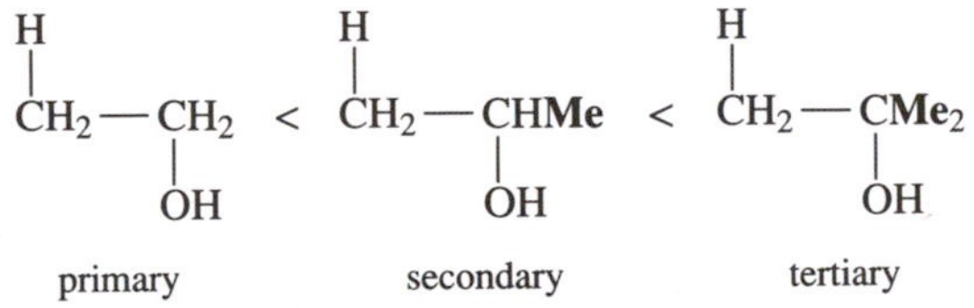

$$
\underset{\text{primary}}{\overset{\overset{\displaystyle H}{|}}{CH_2}-\underset{\underset{\displaystyle OH}{|}}{CH_2}} \quad < \quad \underset{\text{secondary}}{\overset{\overset{\displaystyle H}{|}}{CH_2}-\underset{\underset{\displaystyle OH}{|}}{CHMe}} \quad < \quad \underset{\text{tertiary}}{\overset{\overset{\displaystyle H}{|}}{CH_2}-\underset{\underset{\displaystyle OH}{|}}{CMe_2}}
$$

[9.2] RELATIVE EASE OF DEHYDRATION OF ALCOHOLS

sequence for ease of base-induced elimination of HBr from alkyl bromides in [8.8] (p. 131), when that was occurring *via* the **E1** pathway. This suggests that rapid, reversible protonation of the OH group in the alcohol, by the strong acid H_2SO_4, changes the poor potential leaving group $^\ominus OH$ into the very much better one $:OH_2$:

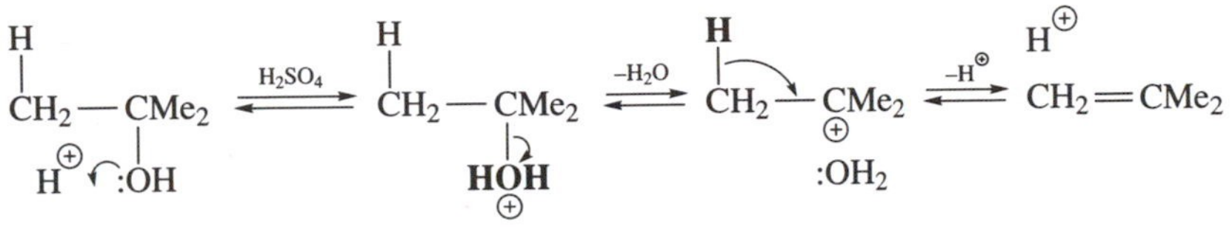

[9.3] E1 PATHWAY FOR DEHYDRATION OF H–CH₂CMe₂—OH

Breaking of the C—O bond, with the departure of H_2O, yields a carbocation intermediate from which $H^\oplus$ is lost to form the product alkene; the driving force for this latter step in the overall reaction stems largely from the stability of the alkene that is thereby produced. Where, in such a carbocation intermediate, there are two differently situated H atoms, as in [9.4], either of them could, in theory, be lost as $H^\oplus$ to complete overall elimination, thereby forming different alkenes:

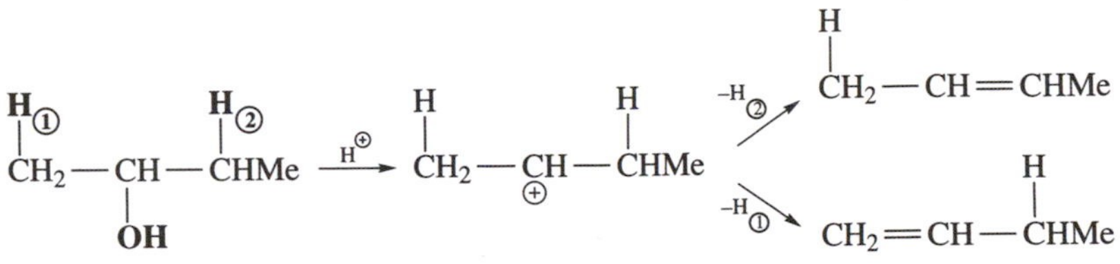

[9.4] ORIENTATION IN E1 ELIMINATION

In practice, it is found that the H atom that is lost preferentially is the one that leads to formation of the alkene with more substituents on its double bond carbon atoms, i.e. the more stable of the two possible alkenes (*cf.* **8.1.6**, p. 137). This preference need not necessarily be absolute, however, and it is not unknown to get a mixture of the two possible alkenes though with one of them predominating.

Although primary alcohols do indeed undergo dehydration with strong acids, such as H_2SO_4, there is no evidence that this reaction proceeds *via* a concerted loss of $H^\oplus$ and H_2O, from first-formed $H—CH_2CH_2—OH_2^\oplus$, i.e.

via an **E2** pathway, comparable to that for base-induced elimination from primary halides (**8.1.1**, p. 128):

$$
\begin{array}{ccccc}
\underset{\substack{| \\ \overset{\oplus}{H}\ \frown :OH}}{\overset{\overset{H}{|}}{CH_2} - CH_2} & \underset{H_2SO_4}{\rightleftharpoons} & \underset{\substack{| \\ \underset{\oplus}{H\overset{..}{O}H}}}{\overset{B:\frown H}{\overset{|}{CH_2}} \overset{\frown}{-} CH_2} & \overset{BH^{\oplus}}{\underset{B:}{\cancel{\longrightarrow}}} & CH_2 = CH_2 \qquad :OH_2
\end{array}
$$

[9.5] LACK OF E2 PATHWAY FOR DEHYDRATION
OF H—CH₂CH₂—OH

This is due largely to there being no adequate base (in the powerfully acid reaction medium) to initiate loss of H$^{\oplus}$ from the β-carbon atom, at the same time as H_2O is lost from the α-carbon, in a single step, concerted pathway.

Given the greater difficulty of forming primary carbocations (compared with tertiary, e.g. that from H—CH₂CMe₂—OH), the necessary breaking of the C—O bond, in the protonated alcohol H—CH₂CH₂—OH$_2^{\oplus}$, must be assisted through solvation of the developing primary cation, H—CH₂CH$_2^{\delta+}$ ··· OH$_2^{\delta+}$, by encircling molecules of solvent. Hardly surprisingly, under the highly acidic conditions employed, there is no evidence of alcohol dehydration proceeding *via* a carbanion intermediate in a two step **E1cB** pathway (*cf.* **8.1.3**, p. 131).

9.1.2 Dehydration of alcohols via ester formation

When, in alcohol dehydration, breaking of the bond to the protonated OH group proves to be difficult, it may be preferable to modify this OH group by means other than protonation in order to improve its ability as a leaving group, e.g. by turning it into an ester. In fact, when CH₃CH₂OH is treated with concentrated H_2SO_4 the most straightforward reaction observed to occur, depending on the conditions, is formation of the ester CH₃CH₂OSO₂OH:

$$
\begin{array}{ccccc}
& \overset{CH_2CH_3}{\underset{}{|}} & & & \\
HOSO_2O \overset{\frown}{-} H \frown :OH & \rightleftharpoons & HOSO_2O^{\ominus} \quad \underset{\substack{| \\ \underset{\oplus}{H\overset{..}{O}H}}}{CH_2CH_3} & \longrightarrow & \underset{ester}{HOSO_2O - CH_2CH_3} \qquad :OH_2
\end{array}
$$

[9.6] FORMATION OF THE ESTER CH₃CH₂OSO₂OH

Overall dehydration to form the alkene could then take place *via* elimination from this ester (reaction 1 in [9.7]),

$$\text{(1)} \quad \overset{\displaystyle H}{\underset{\displaystyle \overset{|}{\text{OSO}_2\text{OH}}}{\text{CH}_2\!-\!\text{CH}_2}} \longrightarrow \overset{\displaystyle H}{\underset{\displaystyle \overset{\ominus}{\text{OSO}_2\text{OH}}}{\text{CH}_2\!-\!\overset{\oplus}{\text{CH}_2}}} \longrightarrow \overset{\displaystyle H^{\oplus}}{\text{CH}_2\!=\!\text{CH}_2}$$

$$\text{(2)} \quad \overset{\displaystyle H}{\underset{\displaystyle \overset{|}{\overset{\oplus}{\text{H}_2\text{O}}}}{\text{CH}_2\!-\!\text{CH}_2}} \longrightarrow \overset{\displaystyle H}{\underset{\displaystyle \text{H}_2\text{O:}}{\text{CH}_2\!-\!\overset{\oplus}{\text{CH}_2}}} \longrightarrow \overset{\displaystyle H^{\oplus}}{\text{CH}\!=\!\text{CH}_2}$$

[9.7] ALKENE FORMATION

as an alternative to (or as well as) *via* elimination from the protonated alcohol (reaction 2 in [9.7]) that we have already considered in **9.1.1** (p. 142).

We also find that reaction between H_2SO_4 and $H—CH_2CH_2—OH$ can, under suitable conditions, lead to the formation of yet another product: the ether, $CH_3CH_2OCH_2CH_3$. This could be produced through reaction of the first formed ester (*cf.* [9.6], p. 144) with another molecule of CH_3CH_2OH, as in reaction 1 in [9.8]:

(1)

$$\underset{\displaystyle \overset{|}{\underset{\displaystyle \text{OSO}_2\text{OH}}{H}}}{\text{CH}_3\text{CH}_2\text{O:}}\;\text{CH}_2\text{CH}_3 \longrightarrow \underset{\displaystyle \overset{|}{\underset{\displaystyle \overset{\ominus}{H}\;\text{OSO}_2\text{OH}}{}}}{\text{CH}_3\text{CH}_2\overset{\oplus}{\text{O}}\!-\!\text{CH}_2\text{CH}_3} \longrightarrow \underset{\displaystyle \text{H}_2\text{SO}_4}{\text{CH}_3\text{CH}_2\text{O}\,-\,\text{CH}_2\text{CH}_3}$$

(2)

$$\underset{\displaystyle \overset{|}{\underset{\displaystyle \overset{\oplus}{\text{OH}_2}}{H}}}{\text{CH}_3\text{CH}_2\text{O:}}\;\text{CH}_2\text{CH}_3 \longrightarrow \underset{\displaystyle H\;\;\text{H}_2\text{O:}}{\text{CH}_3\text{CH}_2\overset{\oplus}{\text{O}}\!-\!\text{CH}_2\text{CH}_3} \longrightarrow \underset{\displaystyle H^{\oplus}}{\text{CH}_3\text{CH}_2\text{O}\,-\,\text{CH}_2\text{CH}_3}$$

[9.8] ETHER FORMATION

This is, of course, a substitution—rather than an elimination—reaction, and the ether could also be obtained by similar attack (reaction 2 in [9.8]) of a molecule of CH_3CH_2OH on the protonated form of the alcohol, $CH_3CH_2OH_2^{\oplus}$, that will also be present in the highly acidic solution. There will be no shortage of CH_3CH_2OH molecules as the reaction—with simple alcohols—is usually carried out in the alcohol as solvent.

Dehydration (or ether formation) *via* initial conversion of the OH group into an ester may, under certain conditions, offer some advantage, in that the necessary ionising ability of the C—O bond—that has to be broken—is thereby increased. The interplay of mechanism in dehydration/ether formation, though much studied, is not entirely clear; though there may be some involvement of an **E2**-like pathway in overall dehydration *via* some esters.

It is possible to effect some degree of control in promoting the formation of either $CH_2{=}CH_2$ or $C_2H_5OC_2H_5$ by manipulation of the reaction conditions. Thus elimination is found to be favoured by higher temperatures, and also by increasing the proportion of acid to that of the alcohol, and *vice-versa* for ether formation. Provided the alcohol is sufficiently volatile, however, the best way of promoting the formation of alkene is by an entirely different route: passing the alcohol vapour over a heated solid catalyst, e.g. Al_2O_3, in a continuous process.

9.2 ELIMINATION TO FORM C≡C?

By analogy with the discussion above, and also with the base-induced elimination of two molecules of HBr from Br—CH_2CH_2—Br to form HC≡CH (**8.2**, p. 138), we might expect acid-catalysed elimination from the 1,2-diol, HO—CH_2CH_2—OH, to yield HC≡CH also. This is not, however, what is actually observed, as elimination of the first molecule of H_2O leads to the formation of an **enol**, CH_2=CH—OH, which isomerises to the more stable carbonyl form, CH_3CHO (*cf.* **5.3**, p. 83):

carbonyl form   enol form

[9.9] DEHYDRATION OF HO—CH_2CH_2—OH

With suitably substituted 1,2-diols it is possible to effect the elimination of two molecules of H_2O, but this is found to result in the formation of a conjugated diene rather than the expected alkyne:

butan - 2,3 - diol   buta - 1,3 - diene (conjugated)

[9.10] DEHYDRATION OF $CH_3CH(OH)CH(OH)CH_3$

No doubt the stabilising effect of conjugation in the developing diene provides the driving force for the reaction to proceed in this direction, rather than towards ketone formation.

9.3 ELIMINATION TO FORM C=O

We have already seen that a number of the nucleophilic addition reactions to carbonyl compounds are reversible (**7.2.1**, p. 105); in the reverse direction these reactions will, of course, be eliminations to form C=O.

9.3.1 Reversal of hemi-acetal, and of acetal, formation

A typical example is reversal of the formation of hemi-acetals (*cf.* **7.2.3**, p. 109), e.g. RCH(OH)OEt:

[9.11] REVERSAL OF HEMI-ACETAL FORMATION

Under normal circumstances, merely trying to isolate a hemi-acetal from solution is in itself sufficient to shift the equilibrium over to such an extent that only the parent aldehyde can actually be obtained.

In contrast to hemi-acetals, acetals themselves (*cf.* **7.2.3**, p. 109), e.g. RCH(OEt)$_2$, are quite stable and may readily be isolated. Their formation from the initial aldehyde (or from the hemi-acetal) requires acid-catalysis under anhydrous conditions, but acetals are very easily hydrolysed back to the parent aldehyde by dilute aqueous acid (i.e. with an excess of H$_2$O present to drive the equilibrium over to the left hand side):

[9.12] HYDROLYSIS OF ACETALS WITH DILUTE AQUEOUS ACID

The stability of acetals, coupled with their ready re-conversion to the parent aldehyde with dilute aqueous acid, has already been mentioned as a method of "protecting" the C=O group (**7.2.3**, p. 110).

9.3.2   Acid-catalysed hydrolysis of esters

We have already considered (7.3, p. 120) the acid-catalysed hydrolysis of esters, e.g. RCO_2Et, as an example of <u>addition</u> to C=O, but—as we saw in [7.40] (p. 122)—although the initial step <u>is</u> addition, the overall reaction also involves an <u>elimination</u>:

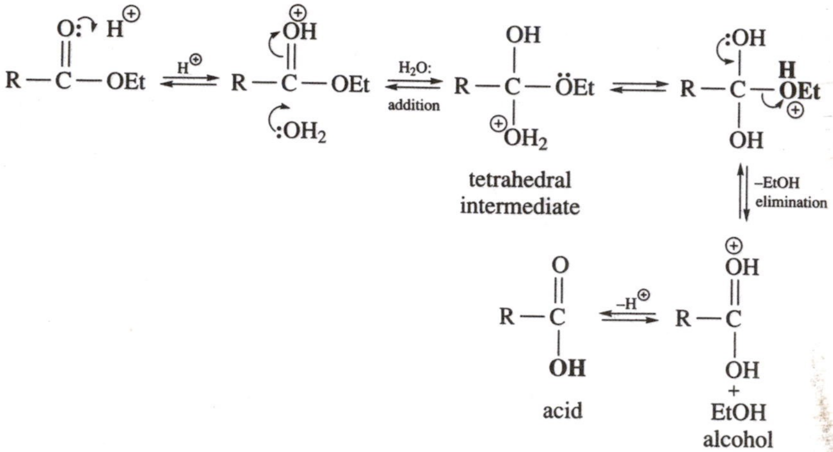

[9.13] ELIMINATION IN ACID-CATALYSED ESTER HYDROLYSIS

Thus initial addition of H_2O: to the protonated ester results in the formation of an intermediate which now has four groups attached to the original carbonyl carbon atom, a <u>tetrahedral</u> intermediate. Proton exchange in this intermediate is followed by elimination of EtOH to yield the protonated form of the acid.

The overall reaction is wholly reversible: it can be made to proceed from left to right—hydrolysis—by using an excess of H_2O, i.e. dilute aqueous acid, or from right to left—ester formation—by using an excess of the alcohol, e.g. EtOH, and a non-aqueous acid catalyst. This is in contrast to base-catalysed hydrolysis of esters which, as we have seen in [7.39] (p. 122), is irreversible. The pathway for acid-catalysed ester formation/hydrolysis shown in [9.13] is much the most common one, but certain structural features in the original ester can prompt a shift to alternative routes.

9.3.2.1   Effect of R′ in RCO₂R′

Thus if the alkyl group in the alcohol end of the ester is capable of forming a relatively stable carbocation, e.g. $-CMe_3 \rightarrow {}^{\oplus}CMe_3$, then this could perhaps be a sufficiently good leaving group to allow its <u>direct</u> elimination from the protonated ester (without requiring initial addition of H_2O as in [9.13], p. 148):

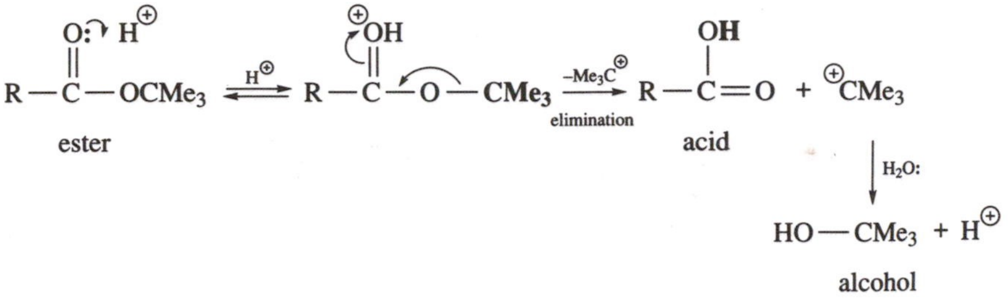

$$R-\overset{\overset{\displaystyle O:\,\,H^{\oplus}}{\|}}{C}-OCMe_3 \underset{}{\overset{H^{\oplus}}{\rightleftharpoons}} R-\overset{\overset{\displaystyle \oplus OH}{\|}}{C}-O-CMe_3 \underset{\text{elimination}}{\overset{-Me_3C^{\oplus}}{\longrightarrow}} R-\overset{\overset{\displaystyle OH}{|}}{C}=O + {}^{\oplus}CMe_3$$

ester   acid

$$\Big\downarrow H_2O:$$

$$HO-CMe_3 + H^{\oplus}$$

alcohol

[9.14] ESTER HYDROLYSIS *VIA* DIRECT ELIMINATION

That this hydrolysis does indeed involve direct elimination of the ester alkyl group—through breaking of its bond to oxygen, and not, as in [9.13] (p. 148) through breaking of this oxygen atom's bond to carbonyl carbon—may be established by carrying out the hydrolysis on the essentially identical ester in which this oxygen atom has been replaced by its heavier isotope, ^{18}O:

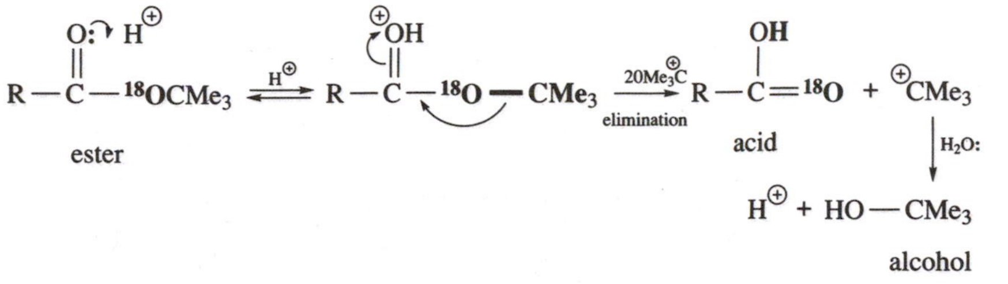

$$R-\overset{\overset{\displaystyle O:\,\,H^{\oplus}}{\|}}{C}-{}^{18}OCMe_3 \underset{}{\overset{H^{\oplus}}{\rightleftharpoons}} R-\overset{\overset{\displaystyle \oplus OH}{\|}}{C}-{}^{18}O-CMe_3 \underset{\text{elimination}}{\overset{20Me_3C^{\oplus}}{\longrightarrow}} R-\overset{\overset{\displaystyle OH}{|}}{C}={}^{18}O + {}^{\oplus}CMe_3$$

ester   acid

$$\Big\downarrow H_2O:$$

$$H^{\oplus} + HO-CMe_3$$

alcohol

[9.15] HYDROLYSIS OF THE ESTER $R-\overset{\overset{\displaystyle O}{\|}}{C}-{}^{18}OCMe_3$

When the two hydrolysis products—acid and alcohol—are isolated at the end of the reaction, it is found that all the ^{18}O isotope is in the acid, and none of it is in the alcohol. This is exactly what we would expect if the hydrolysis had proceeded *via* the direct elimination pathway ([9.14], above), whereas if it had proceeded *via* the tetrahedral intermediate of the usual pathway ([9.13], p. 148) all the ^{18}O would have been found in the alcohol, and none of it in the acid:

[Scheme 9.16]

$$R-\overset{O:}{\underset{\|}{C}}-{}^{18}OCMe_3 \quad \overset{H^{\oplus}}{\rightleftharpoons} \quad R-\overset{\overset{\oplus}{C}OH}{\underset{\|}{C}}-{}^{18}OCMe_3 \quad \overset{H_2O:}{\rightleftharpoons} \quad R-\overset{OH}{\underset{\overset{\oplus}{O}H_2}{C}}-{}^{18}OCMe_3$$

ester   :OH₂

$$R-\overset{O}{\underset{\|}{C}}-OH \quad \overset{-H^{\oplus}}{\rightleftharpoons} \quad R-\overset{\overset{\oplus}{O}H}{\underset{\|}{C}}-OH \quad \underset{elimination}{\overset{-Me_3C^{18}OH}{\rightleftharpoons}} \quad R-\overset{:OH}{\underset{OH}{C}}-{}^{18}\overset{\oplus}{O}CMe_3$$

acid

+
H¹⁸OCMe₃
alcohol

[9.16] PROJECTED HYDROLYSIS OF THE ESTER R—C—¹⁸OCMe₃ BY
THE NORMAL PATHWAY

9.3.2.2   Effect of R in RCO₂R′

Hydrolysis may proceed *via* yet another possible pathway if the alkyl group in the acid moiety of the ester is particularly large and bulky. The hydrolysis of such an ester *via* the usual reaction pathway ([9.13], p. 148) might well be expected to lead to serious crowding, in trying to form the tetrahedral intermediate that would be involved.

Thus hydrolysis of ethyl benzoate, C₆H₅CO₂Et, proceeds readily on heating with dilute acid—no doubt *via* a tetrahedral intermediate;

[Scheme 9.17]

tetrahedral
intermediate

−EtOH
elimination

HOEt

O=C—OH   acid

[9.17] HYDROLYSIS OF ETHYL BENZOATE: C₆H₅CO₂Et

while the broadly similar ester, $2,4,6\text{-}Me_3C_6H_2CO_2Et$ (in which the benzene ring carries bulky Me substituents in both positions *o*- to the CO_2Et group) has still not been hydrolysed after boiling with dilute acid for several days! The usual pathway for hydrolysis ([9.17], p. 150) must thus be inhibited with this ester, presumably by the steric effect of the two *o*-Me groups preventing the formation of what would clearly be an extremely crowded tetrahedral intermediate:

tetrahedral
intermediate

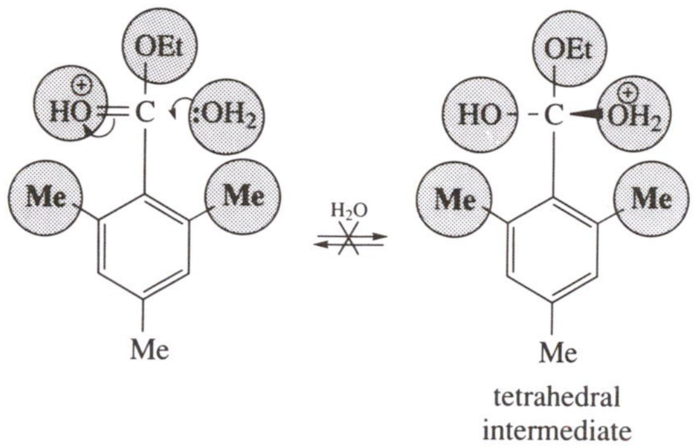

[9.18] CROWDING IN THE TETRAHEDRAL INTERMEDIATE IN THE HYDROLYSIS OF $2,4,6\text{-}Me_3C_6H_2CO_2Et$

If, however, this ester is dissolved in a little concentrated H_2SO_4, and the resulting solution then poured into water, hydrolysis proceeds immediately, and goes to completion! The H_2SO_4 will no doubt protonate the ester on its carbonyl oxygen atom in the usual way (*cf.* [9.17], p. 150), but this cannot lead to hydrolysis as formation of the required tetrahedral intermediate is blocked. There is, however, spectroscopic evidence that protonation may also take place on the otherwise less favoured <u>ester</u> oxygen atom, which <u>can</u> lead to hydrolysis *via* a pathway that is not blocked:

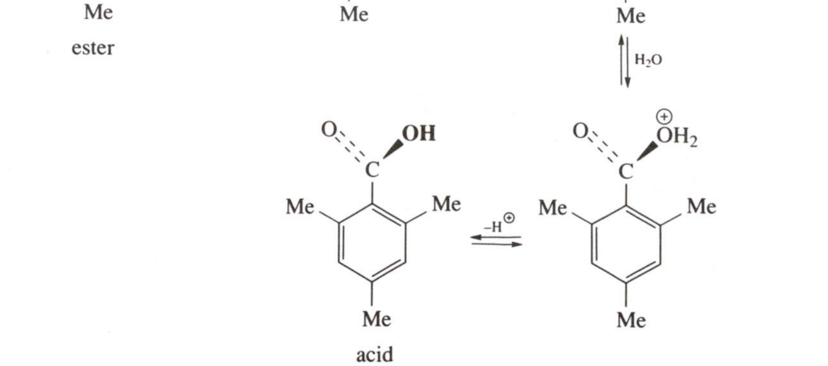

[9.19] HYDROLYSIS OF $2,4,6\text{-}Me_3C_6H_2CO_2Et$

Elimination of EtOH from this species protonated on its ester oxygen atom would lead to an **acyl cation**, which is essentially flat (planar). The approach to this cation of a molecule of H_2O: is now wholly unimpeded— from directions at right angles to the plane of the molecule, i.e. from the front or back of the cation as it is written in [9.19]—and could thus take place readily. The overall reaction could operate in either direction, and it is found that esters can be formed simply by dissolving $2,4,6-Me_3C_6H_2CO_2H$ in concentrated H_2SO_4, and pouring the resulting solution into the appropriate alcohol.

That it is indeed the steric effect of the two o-Me groups that prompts the shift in reaction pathway is borne out by the observation that the isomeric ester, $3,4,5-Me_3C_6H_2CO_2Et$, lacking any substituents in the o-positions,

[9.20] HYDROLYSIS OF $3,4,5-Me_3C_6H_2CO_2Et$

undergoes ready hydrolysis under the usual conditions—the same as those for $C_6H_5CO_2Et$ (*cf.* [9.17], p. 150).

This observation, significant as it is, still does not provide any <u>direct</u> evidence supporting the involvement of an acyl cation as an intermediate in hydrolysis *via* the "hindered" pathway in [9.19]; though it can at least be said that such a species would be stabilised through delocalisation of its $+^{ve}$ charge by the electrons of the adjacent benzene ring (*cf.* [2.13], p. 18). However, on dissolving the highly hindered ester, $2,4,6-Ph_3C_6H_2CO_2Et$ (with C_6H_5 substituents in the two o-positions), in concentrated H_2SO_4, while we obtain a brilliantly coloured solution, on pouring this solution into water in the usual way we fail to obtain any of the expected acid.

The brilliant colour is due to the presence of the tricyclic compound 1,3-diphenylfluorenone (*c.f.* [9.21]), and the formation of this entirely unexpected product can readily be explained by electrophilic attack of the first-formed acyl cation intermediate on one of the o-C_6H_5 substituents, in an <u>internal</u> Friedel–Crafts reaction (*cf.* 3.2.3.2, p. 39). This "trapping" of an acyl cation intermediate does not, of course, prove that hydrolysis of highly hindered esters <u>always</u> proceeds *via* such intermediates, but it does at least make it seem not unreasonable:

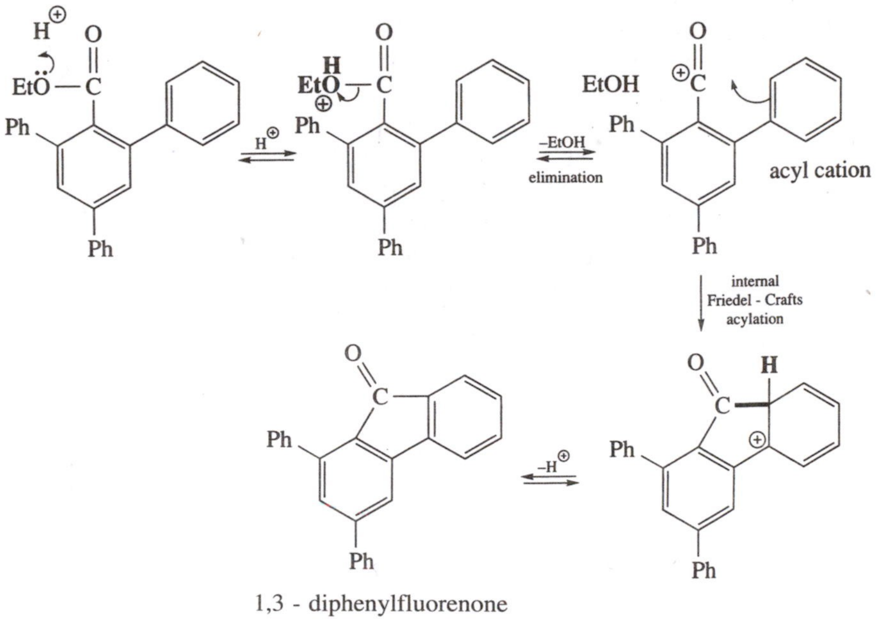

1,3 - diphenylfluorenone

[9.21] "HYDROLYSIS" OF 2,4,6-Ph$_3$C$_6$H$_2$CO$_2$Et

9.4 ELIMINATION TO FORM C=N

We have already seen another example of acid-catalysed elimination, following initial addition, in the reaction of carbonyl compounds with species containing an NH$_2$ group, e.g. hydroxylamine, HONH$_2$ (**7.2.7**, p. 120); a C=N bond is thereby formed:

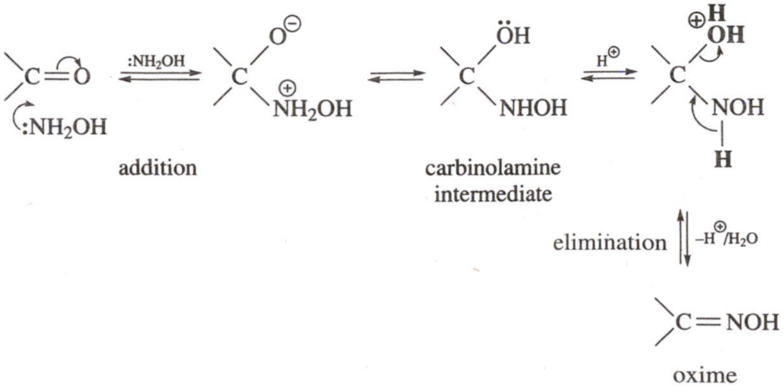

[9.22] ELIMINATION TO FORM C=N IN C=O/HONH$_2$ REACTION

As was mentioned in **7.2.7** (p. 120), the overall reaction is very sensitive to the acidity of the solution in which the reaction is carried out. Thus while

HONH$_2$ is sufficiently nucleophilic to add readily to C=O, a number of other NH$_2$ species are not, e.g. C$_6$H$_5$NHNH$_2$. Such species—before they will react—require initial protonation of the oxygen atom of the C=O group in order to increase the +^{ve} charge on the carbonyl carbon atom, thus making it more responsive to nucleophilic attack. At the same time, too powerful acid-catalysis will result in protonation of the electron pair in :NH$_2$Y, with consequent loss of its nucleophilic ability.

Further, the second stage of the overall reaction—the elimination of H$_2$O from the carbinolamine intermediate—in which we are now primarily interested, also requires acid-catalysis. Hence the three-fold need for careful control of the acidity in order to establish optimum conditions (generally pH 4–5) for the overall reaction.

The involvement of a carbinolamine intermediate may be demonstrated spectroscopically; and also by its actual isolation in the reactions of carbonyl compounds that carry powerful electron-withdrawing substituents, e.g. Cl$_3$CCHO, which are able to stabilise carbinolamines (*cf.* [7.10], p. 108):

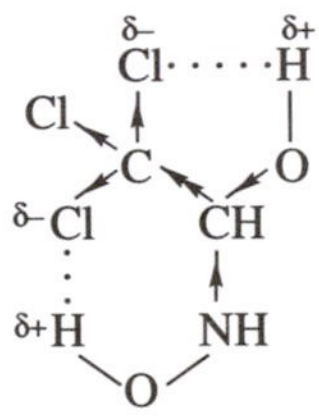

[9.23] ISOLATION OF A CARBINOLAMINE INTERMEDIATE:
Cl$_3$CCH(OH)NHOH

9.5   ELIMINATION TO FORM C≡N

We have already encountered (**7.4**, p. 123) acid-catalysed addition of H$_2$O to the C≡N bond in nitriles, RCN, to form first amides, RCONH$_2$, and then the ammonium salt of the corresponding acids, RCO$_2^{\ominus}$NH$_4^{\oplus}$; this reaction, too, can be reversed. Dehydration of an amide probably proceeds *via* its enol form:

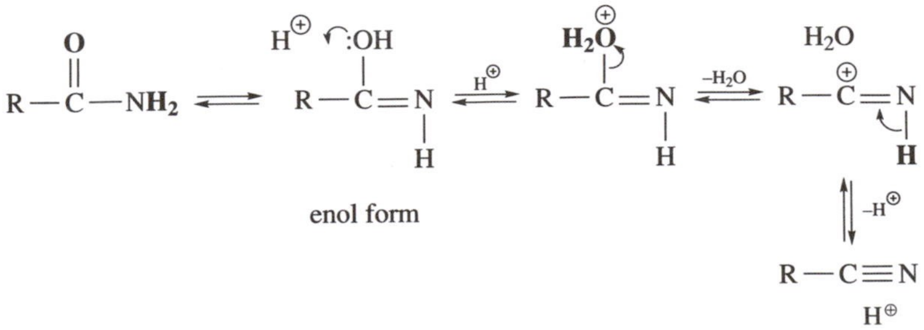

[9.24] DEHYDRATION OF AN AMIDE

The reaction can be induced by concentrated H_2SO_4, but powerful dehydrating agents such as phosphorus pentoxide, P_2O_5, are commonly employed preparatively. These latter reagents probably first form an ester with the OH group of the enol, thereby turning it into a better leaving group.

9.6 SUMMARY

The essential feature of electrophilic elimination is commonly modification, by protonation or other means, of one of the groups being eliminated in order to convert it into a better leaving group. Eliminations to form C=C are much the most common, and a classic example is the dehydration of alcohols, *via* protonation of their OH groups, to form alkenes.

The relative ease of reaction follows the sequence: primary < secondary < tertiary; and proceeds—even with primary alcohols—*via* something approximating to an **E1** pathway. Alternatively, the OH group of the alcohol may be converted into an ester, in order to increase its ability as a leaving group. In either case there may well be competition between elimination to form alkenes, and substitution to form ethers.

Elimination of two molecules of H_2O from 1,2-diols is found to yield not the expected alkynes, but either carbonyl compounds, or conjugated dienes, depending on the overall structure of the diol.

It is possible to reverse a number of the addition reactions of carbonyl compounds to recover the C=O bond, e.g. reversal of acetal formation. More important is the acid catalysed hydrolysis of esters, e.g. RCO_2Et, in which the loss of alcohol from the tetrahedral intermediate is an elimination reaction. Consideration is also given to alternative hydrolysis pathways that are dictated by structural features present in the ester, i.e. the influence of R and R' in RCO_2R'.

Finally mention is made of the formation of C=N, through elimination from the carbinolamine intermediate in oxime formation, etc., and to the formation of C≡N, in the dehydration of amides to nitriles.

10

Radical elimination

Elimination reactions can also be initiated by radicals, but such reactions are of much less importance, or preparative significance, than those initiated by nucleophiles (**8**, p. 127) or by electrophiles (**9**, p. 142). The most common radical-induced elimination reactions are those which result in the formation of C=C:

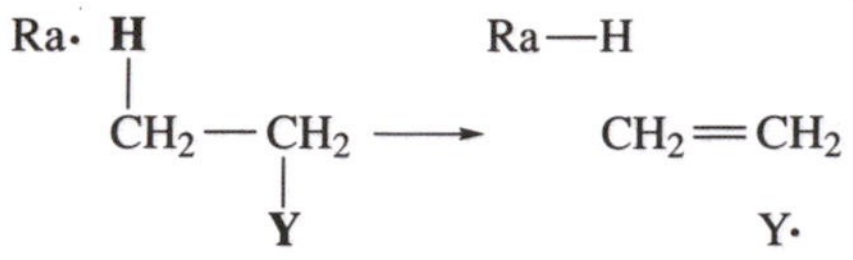

$$\text{Ra·} \quad \overset{\displaystyle H}{\underset{\displaystyle Y}{\overset{|}{\underset{|}{CH_2-CH_2}}}} \longrightarrow \quad \overset{\displaystyle \text{Ra—H}}{\underset{\displaystyle Y·}{CH_2=CH_2}}$$

[10.1] RADICAL INDUCED ELIMINATION

10.1 ELIMINATION TO FORM C=C

The radicals, Ra·, required to initiate elimination may be generated by any of the methods that we have already encountered, e.g. thermolysis ([4.3],

p. 53), photolysis ([4.1], p. 52), or oxidation/reduction ([4.4], p. 53), and a radical pathway for elimination reactions will be promoted by non-polar conditions, i.e. non-polar solvents, and by the absence of manifest electrophiles or nucleophiles.

10.1.1 Possible reaction pathways

In theory at least, radical-induced elimination could proceed *via* pathways analogous to any of those involved in base-induced elimination, i.e. the radical equivalent of **E2** (**8.1.1**, p. 128), **E1** (**8.1.2**, p. 129), or **E1cB** (**8.1.3**, p. 131), depending on the sequence in which the bonds to H and Y are broken.

The possibilities would thus be: (a) a single step pathway in which the H—C and C—Y bonds are broken simultaneously in a concerted process:

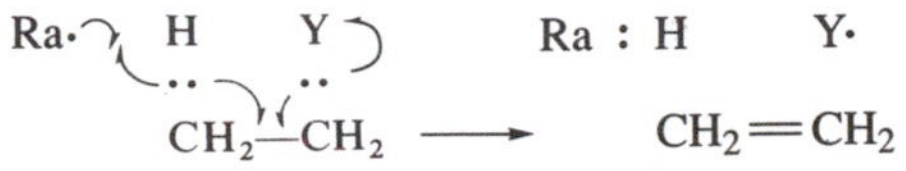

$$Ra\cdot \quad H \quad Y \qquad Ra:H \quad Y\cdot$$
$$CH_2\text{—}CH_2 \longrightarrow CH_2\text{=}CH_2$$

[10.2] SINGLE STEP PATHWAY: H—C AND C—Y BONDS BROKEN SIMULTANEOUSLY

(b) a two step pathway in which the C—Y bond is broken before the H—C bond:

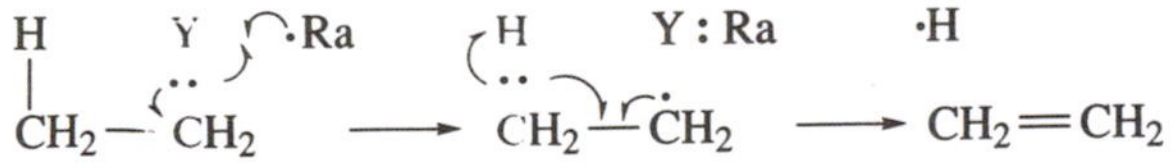

$$H \quad Y \quad \cdot Ra \qquad H \quad Y:Ra \qquad \cdot H$$
$$CH_2\text{—}CH_2 \longrightarrow CH_2\text{—}CH_2 \longrightarrow CH_2\text{=}CH_2$$

[10.3] TWO STEP PATHWAY: C—Y BOND BROKEN BEFORE H—C BOND

(c) a two step pathway in which the H—C bond is broken before the C—Y bond:

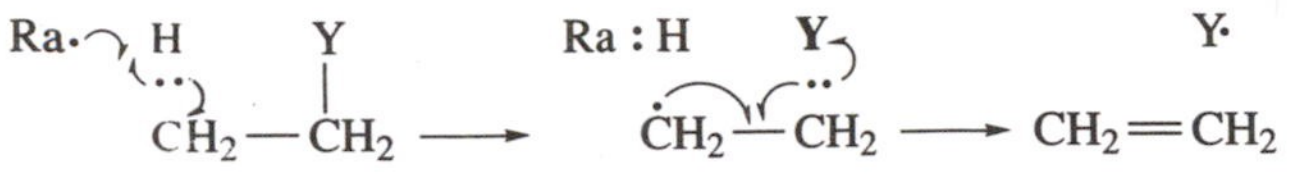

$$Ra\cdot \quad H \quad Y \qquad Ra:H \quad Y \qquad Y\cdot$$
$$CH_2\text{—}CH_2 \longrightarrow \overset{\cdot}{C}H_2\text{—}CH_2 \longrightarrow CH_2\text{=}CH_2$$

[10.4] TWO STEP PATHWAY: H—C BOND BROKEN BEFORE C—Y BOND

At one stage it was believed that elimination of H· and RS· from sulphides, of the form H—CH$_2$—CMe$_2$—SR, proceeded *via* a concerted pathway:

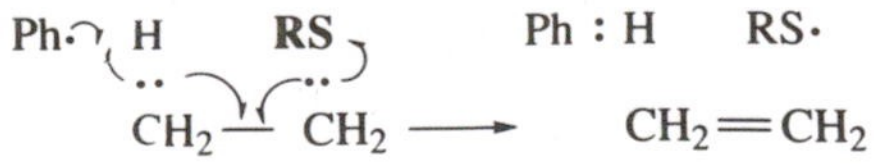

$$Ph \quad H \quad RS \qquad Ph:H \quad RS\cdot$$
$$CH_2\text{—}CH_2 \longrightarrow CH_2\text{=}CH_2$$

[10.5] CONCERTED? LOSS OF H· AND RS· FROM H—CH$_2$CH$_2$—SR

Subsequently, however, it has been shown that elimination of H· and ·SR, from such sulphides, does <u>not</u> occur simultaneously, and there appears to be <u>no</u> C=C forming radical elimination reaction that does indeed involve a one step, concerted pathway. This is, of course, exactly analogous to the reverse reaction—radical-induced addition to C=C—which also proceeds *via* a two step, non-concerted pathway (*cf.* [6.1], p. 87).

There is no real distinction, with radical elimination reactions, between two step pathways (b) and (c) above, in that—after initiation by Ra·—both follow essentially the same course. Whether, in a particular case, it is H or Y that is lost first will be determined purely by whether H· or Y· is abstracted the more easily by the initiator, Ra·. It is found in practice that the atoms most commonly attacked by an initiator radical are H (as in the example in [10.1], p. 156) and halogen (as in the reversal of Br_2 addition to C=C, *cf.* [10.8], p. 159).

10.1.2   Reversal of halogen addition

Reference has already been made (**6.1.1**, p. 88) to the fact that addition of some halogens, e.g. Br_2, to C=C, under radical conditions, is reversible. The reverse reaction—in favour of C=C rather than Hal—C—C—Hal— is found to be promoted by higher temperatures, and by low concentrations of Hal_2.

10.1.2.1   Propene and Cl_2

The effect of higher temperature is seen (*cf.* [4.11], p. 58) in the reaction of Cl_2 with propene, $CH_3CH=CH_2$:

$$CH_2=CH\overset{\centerdot}{C}H_2$$

$$\overset{\centerdot}{C}H_2CH=CH_2 \xrightarrow{Cl:Cl} \overset{\centerdot\centerdot}{Cl}\,CH_2CH=CH_2 + \cdot Cl$$

H·—abstraction   (Cl·)

$$\underset{CH_2CH=CH_2}{\overset{\centerdot\centerdot}{H}}$$

addition   (Cl·)

$$\underset{CH_2\overset{\centerdot}{C}H-CH_2}{\overset{H}{|}}\overset{\centerdot\centerdot}{Cl} \xrightleftharpoons{Cl:Cl} \underset{CH_2CH-CH_2}{\overset{H\ \ Cl}{|\ \ |}}\overset{Cl}{\centerdot\centerdot} + \cdot Cl$$

[10.6] REVERSIBILITY OF Cl_2 ADDITION: $Cl_2/CH_3CH=CH_2$

At lower temperature the expected addition of Cl· (leading to overall addition of Cl_2) is found to take place, but, as the temperature rises, addition of Cl· becomes increasingly reversible, and the alternative of irreversible H· abstraction by Cl· (leading to overall substitution) becomes correspondingly more predominant. Abstraction of H· is also promoted through stabilisation of the developing radical intermediate, involving delocalisation of its unpaired electron by the adjacent double bond (*cf.* [4.11], p. 58).

10.1.2.2 N-bromosuccinimide and cyclohexene

We have also seen the effect of a low concentration of Br_2 (*cf.* [4.13], p. 59) in the use of N-bromosuccinimide to effect the allylic bromination (overall substitution rather than addition) of cyclohexene:

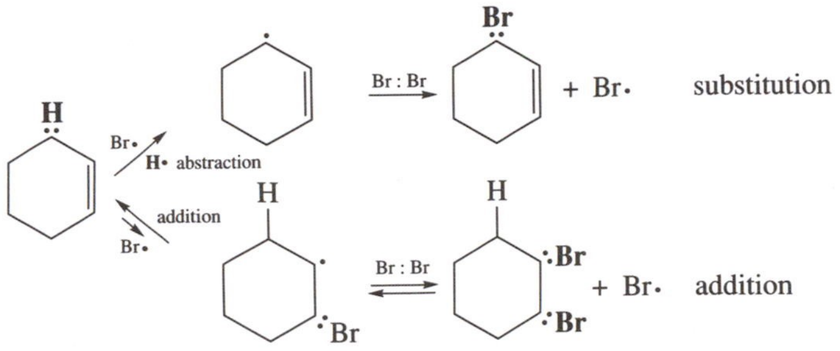

[10.7] REVERSIBILITY OF Br_2 ADDITION: N-BROMOSUCCINIMIDE/
CYCLOHEXENE

Here again there is nothing to stop $Br\cdot$ adding to the double bond, but this is reversible under the conditions of the reaction, whereas abstraction of $H\cdot$ from the allylic CH_2 group by $Br\cdot$ (to form a radical stabilised by the adjacent double bond, *cf.* [10.6], p. 158) is not, and leads to preferential, and irreversible, bromination (overall substitution) at this position.

10.1.2.3 Interconversion of *cis* and *trans* unsaturated compounds

Another useful exploitation of the reversibility of halogen addition is the interconversion of *cis* and *trans* unsaturated compounds, which may be effected by use of a catalytic quantity only of bromine (*cf.* [6.3], p. 88), e.g. with *cis* and *trans* 1,2-diphenylethene in [10.8]:

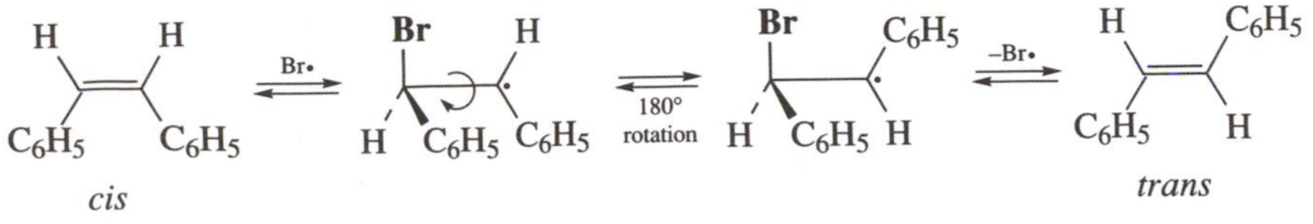

[10.8] REVERSIBILITY OF Br_2 ADDITION: INTERCONVERSION OF
CIS AND *TRANS* 1,2-DIPHENYLETHENE

Such a reaction is commonly carried out in ultraviolet light to produce $Br\cdot$ through photolysis of Br_2.

Starting with the *cis* isomer, as in [10.8], addition of $Br\cdot$ to the double bond yields a radical that can either lose $Br\cdot$—to reform the *cis* starting material—or lose it only after rotation about the now single bond joining the salient carbon atoms—thereby leading to formation of the *trans* isomer. The end result will be an equilibrium mixture of both isomers, the composition of which will reflect their relative stability. In [10.8] the *cis* isomer is likely to be very much the less stable—because its two very bulky C_6H_5 groups are in such close proximity—and the potential "equilibrium mixture" is indeed found to contain only the very much more stable *trans* isomer.

That interconversion does indeed follow such a pathway is borne out by what is found to happen when *cis* 1,2-dibromoethene, BrCH=CHBr, is treated under similar conditions with bromine that is radioactively labelled, *Br_2:

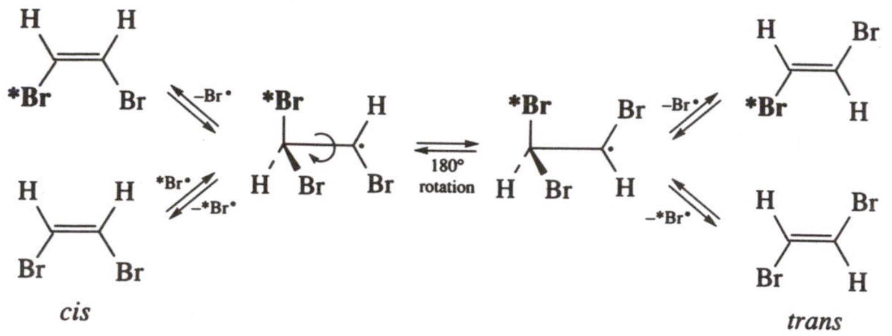

[10.9] REVERSIBILITY OF Br_2 ADDITION: USE OF
RADIOACTIVE *Br_2

Initial addition of *Br· to the *cis* dibromide leads to formation of a radical intermediate in the usual way; this radical intermediate does, however, now have two, distinguishable, bromine atoms, Br and *Br, on its non-radical carbon atom, <u>either</u> of which could be eliminated with equal ease to reform C=C.

Loss of *Br· from this radical intermediate would result in reformation of the original *cis* dibromide, while loss of Br· would result in formation of the analogous *cis* dibromide containing a radioactive bromine atom, *Br. If, however, loss of a bromine radical takes place only <u>after</u> rotation about the now single bond joining the salient carbon atoms, the result will be formation of the *trans* dibromide: loss of *Br· would result in formation of the *trans* isomer of the original dibromide, while loss of Br· would result in formation of the analogous *trans* dibromide containing a radioactive bromine atom, *Br.

If interconversion follows the pathway suggested in [10.9]—and in [10.8]—we would expect that <u>both</u> *cis* and *trans* dibromides (after isolation from the equilibrium mixture) will be found to contain radioactive bromine, *Br: which is exactly what we <u>do</u> find!

10.2 FRAGMENTATION REACTIONS

The reaction that we have just been considering, involving elimination of a halogen atom from a radical to form a carbon–carbon double bond, is often referred to as a **fragmentation** reaction, and such reactions may involve the elimination of atoms or groups other than halogen, e.g. thiyl radicals, RS·.

10.2.1   Thiyl radicals RS·

Thus the interconversion of *cis* and *trans* 1,2-diphenylethene (*cf.* [10.8], p. 159) may also be effected by catalytic quantities of thiols, RSH, which readily yield thiyl radicals, RS·:

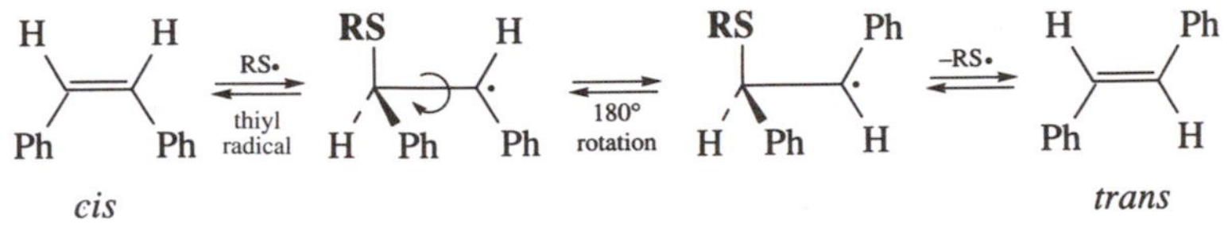

[10.10] FRAGMENTATION INVOLVING THIYL RADICALS: RS·

Fragmentation reactions of this kind are often referred to as β-scission reactions, because the atom or group being eliminated is lost from the carbon atom β- to the radical centre.

10.2.2   Alkoxy radicals RR'R"CO·

A number of other fragmentation reactions are known as well as the reversal of radical addition to C=C; a good example is β-scission of alkoxy radicals, RR'R"CO·, e.g. Me₃CO·:

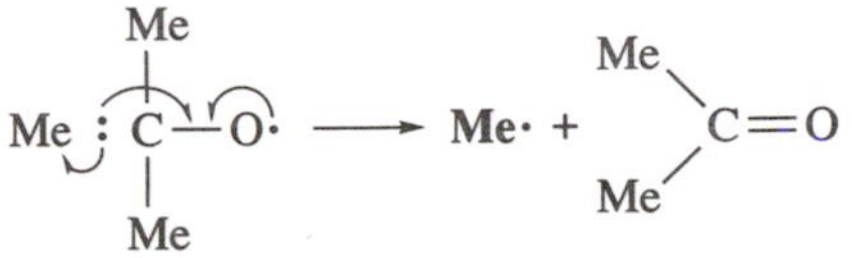

[10.11] FRAGMENTATION OF ALKOXY RADICALS: Me₃CO·

The initial alkoxy radical, Me₃CO· (which may be generated in a number of different ways, e.g. thermolysis of Me₃CO—Cl, or Me₃CO—OCMe₃) loses a methyl radical, Me·, thereby forming a C=O bond in the product, propanone; the driving force of the reaction no doubt stems in part from the strong C=O bond that is being formed. Such fragmentation reactions are found to be favoured by higher temperature, and also by the presence in R₃CO· of alkyl groups which will form somewhat more stable alkyl radicals, R·.

This latter effect is reflected in the preferential elimination that occurs when the alkyl groups in R₃CO· are different from each other. It is then found that a secondary radical is eliminated ≈ 50 times more readily than is a primary, while a tertiary radical is eliminated ≈ 300 times more readily. Thus in [10.12],

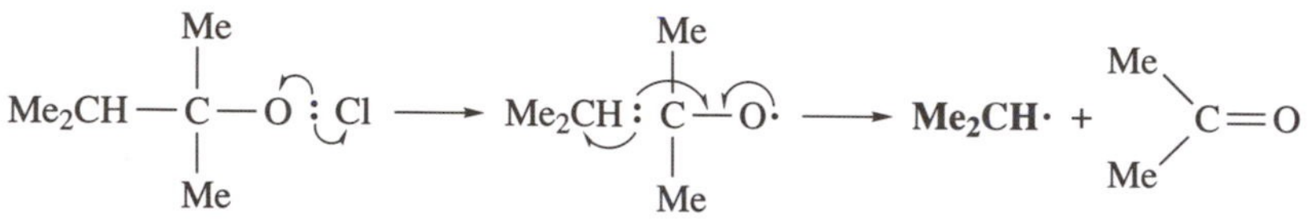

[10.12] PREFERENTIAL ELIMINATION FROM RR′R″CO·

it is the more stable potential radical, $Me_2CH\cdot$, which is eliminated preferentially.

10.2.3   Acyl radicals $R\dot{C}{=}O$

Aliphatic (but not aromatic) aldehydes, RCHO, may be induced to eliminate CO—**decarbonylation**—under the influence of heat or light, provided alkyl peroxides, RO—OR are present:

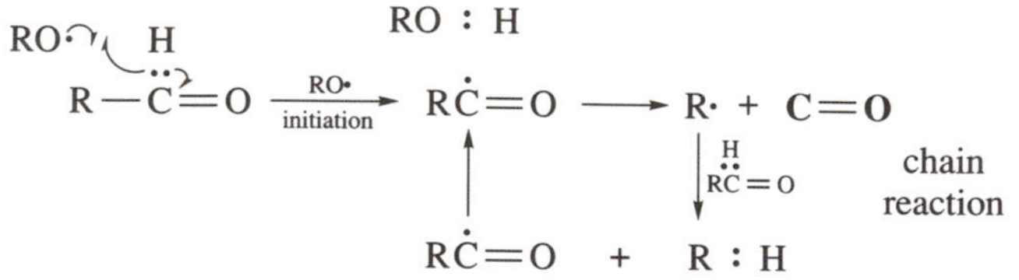

[10.13] DECARBONYLATION OF ALDEHYDES: RCHO

Heat/light thermolyses/photolyses the peroxide to yield alkoxy initiator radicals, $RO\cdot$, which are able to abstract $H\cdot$ from the aldehyde, RCHO, to form an acyl radical, $R{-}\dot{C}{=}O$. Under the reaction conditions, this radical then eliminates CO, forming an alkyl radical, $R\cdot$, which can abstract $H\cdot$ from a further molecule of RCHO to set up an on-going chain reaction. In common with other fragmentation reactions, the more stable $R\cdot$ is the more readily will the reaction proceed.

10.2.4   Acyloxy radicals $RCO_2\cdot$

We have already seen an example of such a fragmentation reaction in the fission of benzoyl peroxide, $PhCO_2{-}O_2CPh$ ([4.19], p. 62), to yield benzoyloxy radicals, $PhCO_2\cdot$, which **decarboxylate** at quite low temperature (80°C) to form phenyl radicals, $Ph\cdot$:

$$\underset{\text{Ph}-\overset{\overset{\text{O}}{\|}}{\text{C}}-\text{O}:\text{O}-\overset{\overset{\text{O}}{\|}}{\text{C}}-\text{Ph}}{} \longrightarrow \text{Ph}-\overset{\overset{\text{O}}{\|}}{\text{C}}-\text{O}\cdot + \cdot\text{O}-\overset{\overset{\text{O}}{\|}}{\text{C}}-\text{Ph} \qquad \text{peroxide fission}$$

$$\text{Ph}:\overset{\overset{\text{O}}{\|}}{\text{C}}\text{—O}\cdot \longrightarrow \text{Ph}\cdot + \text{O}{=}\text{C}{=}\text{O} \qquad \text{decarboxylation}$$

[10.14] DECARBOXYLATION OF ACYLOXY RADICALS: PhCO$_2$·

The O—O bond in acyl peroxides is a very weak one, and breaks extremely easily; indeed more vigorous conditions are required for decarboxylation of the resulting acyloxy radical, than for fission of the initial acyl peroxide. Decarboxylation of aliphatic acyloxy radicals, RCO$_2$·, is found to proceed even more readily than that of the aromatic variety, ArCO$_2$·.

Whereas the decarboxylation of PhCO$_2$· is no more than a convenient way of producing phenyl radicals, Ph·, there is a radical decarboxylation reaction that can, in suitable cases, be used preparatively—the **Hunsdiecker** reaction:

$$\text{RCO}_2\text{Ag} \xrightarrow{\text{Br}_2} \text{RBr} + \text{CO}_2 + \text{AgBr}$$

[10.15] HUNSDIECKER REACTION

This involves reaction of the silver salt of a carboxylic acid with bromine, and results overall in loss of CO$_2$ to form the corresponding alkyl (or aryl) bromide. The first step of the reaction—the formation of an acyl hypobromite, RCO$_2$Br—does not involve radicals, but the subsequent fragmentation of this species does:

$$\text{RCO}_2\text{Ag} + \text{Br}_2 \longrightarrow \text{RCO}_2\text{Br} + \text{AgBr} \qquad \text{non - radical reaction}$$

$$\text{RCO}_2 : \text{Br} \longrightarrow \text{RCO}_2\cdot + \cdot\text{Br} \qquad \text{initiation}$$

$$\text{R}:\overset{\overset{\text{O}}{\|}}{\text{C}}\text{—O}\cdot \longrightarrow \text{R}\cdot + \text{O}{=}\text{C}{=}\text{O}$$

chain reaction

$$\text{R}:\overset{\overset{\text{O}}{\|}}{\text{C}}-\text{O}\cdot + \text{R}:\text{Br}$$

[10.16] PATHWAY FOR HUNSDIECKER REACTION

Easy fission of the weak O—Br bond in a few molecules of the acyl hypobromite provides enough RCO$_2$· radicals to initiate an on-going chain reaction leading to the formation of RBr and CO$_2$.

It should be remembered that decarboxylation of suitable carboxylic acid derivatives can also be effected under base-catalysed conditions (*cf.* [8.17], p. 135).

10.2.5   Depolymerisation

Reference has already been made to the radical-initiated polymerisation of alkenes (**6.1.3**, p. 90), in which successive additions of monomer to the growing polymer radical constitute the rapid propagation step of the overall chain reaction. It is, however, found that if the conditions are varied—particularly if the temperature is raised sufficiently—this step may be reversed, leading to successive elimination of monomer molecules from the polymer radical in a depropagation step, leading to overall depolymerisation. The effect of temperature is such that there is often a ceiling temperature, above which the formation of long chain polymers is no longer possible.

Thus poly methyl methacrylate (perspex), $[CH_2—C(Me)CO_2Me]_n$, is

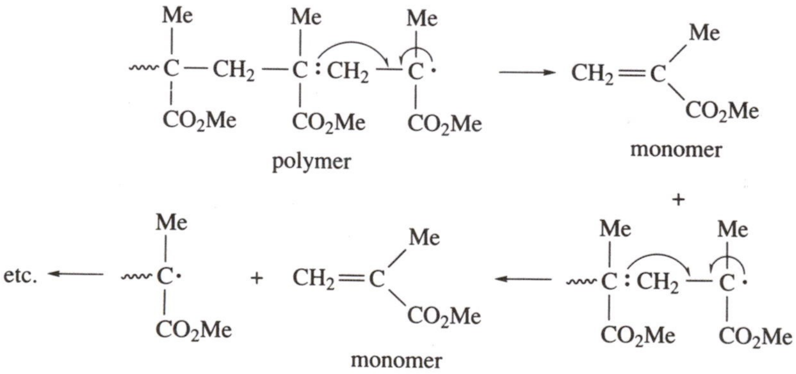

[10.17] DEPOLYMERISATION OF POLY METHYL METHACRYLATE

found to have a ceiling temperature of 190°C: it may indeed be depolymerised to monomer ($CH_2=C(Me)CO_2Me$), in high yield, simply by heating the polymer in an open vessel over a flame. Depolymerisation may be effected at even lower temperature (130°C) if the polymer is, at the same time, irradiated with light of suitable wavelength; this response to heat and light does serve to restrict the possible uses of poly methyl methacrylate as a polymeric material.

10.2.6   Azoalkanes R—N=N—R

The thermal fission of azoalkanes, R—N=N—R, results in the formation

$$R : N=N : R \longrightarrow R\cdot + N\equiv N + \cdot R$$

[10.18] THERMAL FISSION OF AZOALKANES

of a molecule of nitrogen, $N{\equiv}N$, which, being perhaps the most effective leaving group there is, no doubt supplies the driving force for fission of the strong C—N bonds.

This may explain why, unlike elimination to form C=C (**10.1.1**, p. 157) which is apparently <u>never</u> a concerted process, both R· radicals are often eliminated simultaneously from azo compounds in a single concerted step; that is unless one R· is very much more stable than the other, in which case it will be lost first, and a two step pathway will result. The ease with which azoalkane fragmentation takes place is determined very largely by the relative stability of the radicals, R·, that result. Thus whereas fission of MeN=NMe requires a temperature of $\approx 400°C$, $Ph_2CHN{=}NCHPh_2$ is found to decompose readily at 65°C.

A particular feature of azoalkane fragmentation is that it constitutes an easy method for the generation of radicals *in situ* to initiate other radical processes, e.g. polymerisation. One particularly useful example is the thermal fission of "azoisobutyronitrile" (AIBN), $Me_2C(CN)N{=}NC(CN)Me_2$, to form the very useful initiator radical, $Me_2C(CN)·$:

CN   CN   CN   CN
Me₂C : N≡N : CMe₂ ⟶ Me₂C· + N≡N + ·CMe₂

[10.19] GENERATION OF Me₂C(CN)·

The fission of this azo compound occurs readily at quite low temperature because of the stabilisation of the forming radical, $Me_2C(CN)·$, that results from delocalisation of its unpaired electron by the $C{\equiv}N$ substituent. Apart from the ease with which they may be generated, the relative stability of $Me_2C(CN)·$ radicals means that their initiation of other radical reactions, e.g. vinyl polymerisation (**6.1.3**, p. 90), is generally slower, and more controlled, than with many other potential initiators, e.g. peroxides.

10.3 DISPROPORTIONATION REACTIONS

Reference has already been made ([6.8], p. 91) to the termination of individual reaction chains in radical reactions through the collision of two radicals with each other, e.g. in vinyl polymerisation:

RO(CH₂)ₙCH₂CH₂ ⌒⌣ CH₂CH₂(CH₂)ₙOR → RO(CH₂)ₙCH₂CH₂ : CH₂CH₂(CH₂)ₙOR

[10.20] TERMINATION OF RADICAL REACTION CHAINS:
DIMERISATION

Termination is shown here as involving the pairing of two electrons, one from each radical, to form a bond between them—**dimerisation**.

This is not, however, the only way in which the two radicals can react with each other in chain termination; alternatively—in the commonest of all radical reactions—one radical can abstract H· from the other, in this case from the carbon atom β- to the one carrying the unpaired electron (β-scission):

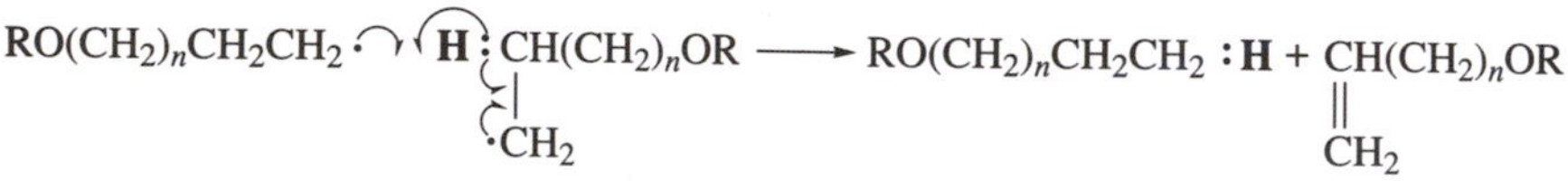

$$RO(CH_2)_nCH_2CH_2 \frown H \vdots CH(CH_2)_nOR \longrightarrow RO(CH_2)_nCH_2CH_2 \vdots H + CH(CH_2)_nOR$$
$$\overset{\downarrow}{\underset{\cdot CH_2}{C}} \qquad\qquad\qquad\qquad \overset{\|}{CH_2}$$

[10.21] TERMINATION OF RADICAL REACTION CHAINS: DISPROPORTIONATION

The overall result—one alkyl radical being converted into an alk*ane*, and the other into an alk*ene*—is known as **disproportionation**, because one radical has gained an atom of hydrogen (addition), while the other radical has lost one (elimination).

Disproportionation between radicals is not, however, confined to termination of radical chains in vinyl polymerisation. Thus we have already seen a different kind of example ([4.21], p. 63) in the course of phenylation: the reaction of aromatic compounds with C_6H_5·. Disproportionation tends to occur, at the expense of simple dimerisation, in situations where the latter reaction is likely to be impeded by steric factors.

Thus the interaction of Me_3C· radicals is found to lead to considerably more disproportionation than dimerisation:

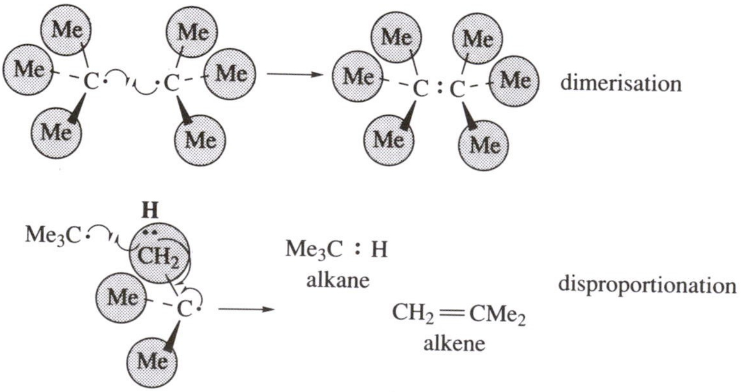

dimerisation

$Me_3C \vdots H$
alkane

$CH_2 = CMe_2$
alkene

disproportionation

[10.22] INTERACTION OF Me_3C· RADICALS

This happens because pairing of the electrons on the two central carbon atoms of Me_3C· radicals (dimerisation) will be greatly hindered by the bulky Me substituents, whereas abstraction of H· from the periphery of one radical by the other (disproportionation) will not.

10.4 SUMMARY

Elimination initiated by radicals, to form C=C, is much rarer than elimination induced by nucleophiles or electrophiles, and is of correspondingly less preparative significance. Radical-induced elimination proceeds by a two step pathway, in which abstraction of one atom or group (very often H or Hal) by an initiator radical is followed by loss of the other, as a radical, from the β-carbon atom (β-scission).

Radical addition reactions—particularly of halogens—are often readily reversible, and the resultant elimination can be exploited, e.g. in the chlorination of $CH_3CH=CH_2$ by use of higher temperature; in the bromination of cyclohexene by use of a low concentration of Br_2 (N-bromosuccinimide); and in the interconversion of *cis* and *trans* unsaturated compounds.

Overall elimination is generally completed by loss of a radical from the intermediate that was produced by initial abstraction of an atom or group by an initiator: this process is called a **fragmentation** reaction. Examples of such reactions include fragmentation of $RS\cdot$, $RR'R''CO\cdot$, $R\dot{C}=O$, $RCO_2\cdot$, $RO(CH_2)_nCH_2CH_2\cdot$ in reversal of the propagation step of radical polymerisation (*cf.* [6.8], p. 91); and also thermolysis of R—N=N—R.

Finally, there is a consideration of **disproportionation** reactions in which two radicals are reacting with each other, e.g. in the termination of reaction chains in vinyl polymerisation. As an alternative to pairing their unpaired electrons to form a bond between them (dimerisation), in disproportionation one radical abstracts H· from the β-carbon atom of the other thereby becoming an alkane, while the second radical thus becomes an alkene.

Index

Notes

Notes

Notes